WANGLUOHULIAN YU LUYOUJISHU SHIYAN ZHIDAOSHU

网络互联与路由技术实验指导书

叶 涛 刘 昕 昝风彪／编著

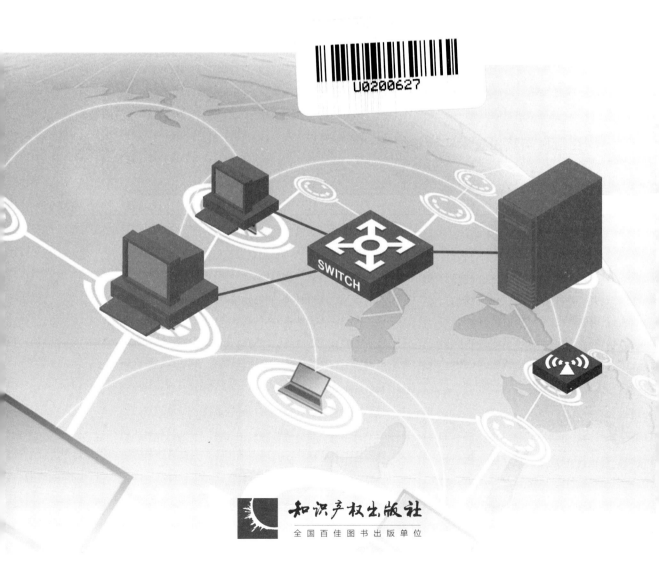

知识产权出版社
全国百佳图书出版单位

图书在版编目（CIP）数据

网络互联与路由技术实验指导书 / 叶涛，刘昕，昝风彪编著.—北京：知识产权出版社，2017.11
ISBN 978-7-5130-5079-1

Ⅰ.①网… Ⅱ.①叶… ②刘… ③昝… Ⅲ.①互联网络－实验－高等学校－教材②路由器－实验－高等学校－教材 Ⅳ.①TP393.4-33②TN915.05-33

中国版本图书馆CIP数据核字（2017）第202402号

内容提要

本实验指导书以杭州华三通信技术有限公司（简称H3C）MSR 20系列、30-20、50-60路由器和S3600、S3610交换机等网络设备为实验硬件平台，以H3C COMWARE v3/v5操作系统为软件环境，提供了常用网络设备管理基本操作、网络设备登录及认证、网络设备系统软件升级、交换机端口配置与转发表维护、STP协议配置、VLAN划分配置、IP路由基础及VLAN间路由配置、静态路由配置、RIP协议配置、OSPF协议配置、PPP协议配置、帧中继配置、NAT地址转换配置、ACL访问控制列表配置、以太网交换机端口安全配置等一系列实验操作。

本书适合网络工程、计算机科学与技术等专业的高年级学生、网络实验指导教师及网络工程技术人员阅读，也可作为H3CNE认证培训实验指导教材。

责任编辑：田　姝　彭喜英　　　　　　　　　　　　责任出版：孙婷婷

网络互联与路由技术实验指导书

叶　涛　刘　昕　昝风彪　编著

出版发行：知识产权出版社有限责任公司	网　址：http://www.ipph.cn
电　话：010－82004826	http://www.laichushu.com`
社　址：北京市海淀区气象路50号院	邮　编：100081
责编电话：010-82000860转8539	责编邮箱：pengxyjane@163.com
发行电话：010-82000860转8101	发行传真：010-82000893
印　刷：北京嘉恒彩色印刷有限责任公司	经　销：各大网上书店、新华书店及相关专业书店
开　本：787mm×1092mm　1/16	印　张：11.5
版　次：2017年11月第1版	印　次：2017年11月第1次印刷
字　数：282千字	定　价：35.00元

ISBN 978-7-5130-5079-1

出版权专有　侵权必究
如有印装质量问题，本社负责调换。

本书常用 H3C 网络设备图标

路由器	IPv6路由器	便携计算机	计算机	服务器	服务器群
集线器	网桥	交换机	三层交换机	二层交换机	AP
帧中继交换机	运营商传输设备	CSU/DSU	Modem	Access server	Firewall
移动人员	用户	用户群	网络云		

前　言

"网络互联与路由技术"是网络工程专业的一门核心课程。它是在掌握网络基本知识的基础上，对网络互联概念、互联方法、网络互联设备工作原理及其管理配置操作进行学习。该课程的实践操作是学习和研究网络互联技术的重要环节，是其理论课程的必要补充。其目的是通过一系列由浅入深的实验实践帮助学习者理解网络互联相关概念，掌握网络设备管理配置操作和调试诊断方法，培养构建和维护中小型企业网络的综合实践能力。

作者依据多年来对本科学生进行网络教学和相关科研实践的经验，在征求了网络工程专业相关教师、高年级学生及企业网络工程技术人员意见的基础上，从工程实践和应用角度出发，完成了本实验指导书的编写。

本实验指导书以 H3C MSR 20 系列、30-20、50-60 路由器和 S3600、S3610 交换机等网络设备为硬件平台，以 H3C COMWARE v3/v5 操作系统为软件环境，根据学习循序渐进的认知顺序和工程实际应用需求，按实验目的、实验内容、实验设备与组网图、实验相关知识、实验过程、实验思考六个环节组织了每一个实验方案。方案侧重于网络互联技术和网络设备管理配置操作方法介绍，也强调在实践操作中进一步促进学习者对相关理论知识的理解和应用。

全书共提供了十六个教学实验和两个综合扩展实验，适合作为网络工程、计算机科学与技术等专业网络互联技术相关课程的实验指导教材。实验一～实验三实践了网络设备管理与维护的基本操作；实验四～实验七实践了交换机管理与配置操作；实验八～实验十一实践了路由器管理与配置操作；实验十二、实验十三实践了广域网常用协议配置操作；实验十四～实验十六实践了网络设备的安全设置操作；综合实验一、综合实验二分别介绍了网络设备调试诊断操作和网络互联技术综合应用实践，是教学实验所需知识的进一步补充和扩展。

本实验指导书由叶涛主编，负责全书统稿修订工作。该书编写具体分工为：叶涛负责撰写了实验一～实验十一；刘昕负责撰写了实验十二～实验十六；昝风彪负责撰写了综合实验一、综合实验二，并对全书进行了审阅修订；刘昕指导徐超、李竞阳、娄甲甲、曹有林等学生绘制了全书插图，进行了阅读校对和实验操作验证，提出了诸多参考意见。

虽然作者在本书编写过程中力求叙述准确、完善，鉴于计算机网络技术发展迅速，作者水平和时间有限，书中难免存在不妥之处，恳请同行专家和读者批评指正。

在此，我们衷心感谢为本实验指导书出版做出贡献的组织、企业及个人，他们以不同的方式为本书编写做出了重要贡献。本书在编写过程中获得作者单位的支持和其他同事的帮助，对编写本书时所参考书籍的作者在此一并表示诚挚的感谢。也感谢杭州华三通信技术有限公司授权本书使用 H3C 设备相关图标及操作命令。

编 者

2017 年 8 月 22 日于青海民族大学

目 录

实验一　常用网络设备管理基本操作 ………………………………………………………… 1
实验二　网络设备登录及认证管理 …………………………………………………………… 19
实验三　网络设备系统软件升级 ……………………………………………………………… 31
实验四　交换机端口设置与转发表维护 ……………………………………………………… 40
实验五　链路聚合配置 ………………………………………………………………………… 49
实验六　STP 协议配置 ………………………………………………………………………… 55
实验七　VLAN 配置 …………………………………………………………………………… 61
实验八　IP 路由基础及 VLAN 间通信 ……………………………………………………… 71
实验九　静态路由配置 ………………………………………………………………………… 87
实验十　RIP 协议配置 ………………………………………………………………………… 94
实验十一　OSPF 协议配置 …………………………………………………………………… 106
实验十二　PPP 配置 …………………………………………………………………………… 120
实验十三　帧中继配置 ………………………………………………………………………… 128
实验十四　ACL 包过滤 ………………………………………………………………………… 137
实验十五　NAT 配置 …………………………………………………………………………… 144
实验十六　交换机端口安全技术 ……………………………………………………………… 160
综合实验一　网络设备调试与诊断操作 ……………………………………………………… 164
综合实验二　网络互联技术综合应用实践 …………………………………………………… 173
参考文献 ………………………………………………………………………………………… 175

实验一 常用网络设备管理基本操作

一、实验目的

① 理解用户界面、用户界面编号、用户界面视图、用户视图等相关概念；
② 掌握 PC 机与网络设备 Console 口连接方法；
③ 熟悉超级终端和 SecureCRT 软件的基本操作；
④ 熟悉设备名称修改、系统时间修改、快捷键设置等常用系统维护命令；
⑤ 掌握查看设备版本、当前配置、接口信息等操作方法；
⑥ 掌握网络设备文件、目录管理操作。

二、实验内容

① 使用专用配置线、COM 口转 USB 转接器实现 PC 与网络设备 Console 口的物理连接；
② 安装超级终端或 SecureCRT 软件，配置相应参数登录网络设备；
③ 用 quit、<Ctrl+Z>、return、system-veiw 等操作命令实现不同视图间的转换；
④ 使用 sysname、clock 等命令维护设备系统配置；
⑤ 使用 display 命令查看设备版本、当前配置、接口信息；
⑥ 使用 dir、mkdir、rmdir、more 等命令管理网络设备文件和目录。

三、实验设备与组网图

1．实验设备

一台 H3C MSR30-20 路由器,一台计算机，超级终端或 Secure CRT 软件，一条 Console 专用配置线缆，COM 口转 USB 转接器，一台 H3C S3610 交换机。

2．组网图

常用网络设备管理基本操作实验组网如图 1-1 所示。

图 1-1 实验组网

四、实验相关知识

1．用户界面

交换机、路由器等网络互联设备类似一台专用计算机，需通过监视器来监控、管理设备运行情况。当网络设备开机运行后，用户需通过某种接口或界面登录设备来监控、管理设备运行状态及用户与网络设备之间的会话过程，H3C 系统中将这种监控、管理用户与设备之间的会话接口称为用户界面。

（1）H3C 系统支持的用户界面

H3C 系列路由器与高端以太网交换机支持四种用户界面类型，分别是 CON 控制端口用户界面、AUX 辅助端口用户界面、VTY 虚拟类型终端接口用户界面、TTY 实体类型终端用户界面。H3C 系列中低以太网交换机只支持 CON 用户界面和 VTY 用户界面。

CON 用户界面：Console 端口是一种由网络设备主控板提供的串行物理接口，通过设备供应商提供的专用配置线，用户终端的串行接口可与网络设备 Console 口直接连接，实现对网络设备的本地登录。CON 用户界面专门用来管理和监控通过 Console 口登录的用户。H3C S3610 系列以太网交换机 AUX 用户界面是控制端口用户界面。

AUX 用户界面：路由器提供一个物理 AUX 口，是一种线设备端口，可以如 Console 口一样进行本地配置。AUX 口还可以进行远程配置，通常用于通过 Modem 进行拨号访问，端口类型为 EIA/TIA-232 DTE。

VTY 用户界面：网络设备的 Ethernet、serial 等接口工作在同步方式下，通过 Telnet 或 SSH 服务建立的一种虚拟类型终端接口，称为 VTY 虚拟线路，用来管理和监控通过 VTY 方式登录的用户。VTY 口属于逻辑终端线。

TTY 用户界面：指路由器异步串口或同/异步串口，如 Serial 接口工作在异步方式下，采用专用网线与用户终端连线，通过 Modem 拨号访问建立连接的一种实体类型终端接口，用来管理和监控通过 TTY 方式登录的用户。

（2）H3C 系统用户界面编号

交换机、路由器等网络设备都可同时登录多个用户，一个用户登录时将占用一个用户界面。为区分和引用不同用户界面，H3C 系统给用户界面分配了用户界面编号，编号方式有两种：绝对编号方式和相对编号方式。

1）绝对编号方式

绝对编号方式是按设备支持的所有类型用户界面统一编号，它可以唯一地指定一个用户界面或一组用户界面。绝对编号从 0 开始自动编号，每次增大 1，首先给所有 Console 用户界面编号，其次是所有 TTY 用户界面，再次是所有 AUX 用户界面，最后是所有 VTY 用户界面。使用 display user-interface 命令可查看设备当前支持的用户界面及它们的绝对编号。

2）相对编号方式

相对编号是按各种类型用户界面的独立自编号，该方式只能指定某种类型的用户界面中的一个或一组，而不能跨类型操作。

相对编号方式的形式是："用户界面类型编号"，遵守规则如下。

控制台的编号：CON 0；

辅助线的编号：AUX 0；

TTY 的编号：第一条为 TTY 0，第二条为 TTY 1，依此类推；

VTY 的编号：第一条为 VTY 0，第二条为 VTY 1，依此类推。

2．视图

视图可以看作一组功能相关命令的集合，是用户与系统会话操作界面。为便于用户安全使用 H3C 系统提供的丰富配置管理和查询命令，系统将所有命令按功能进行分类组织。视图采用分层结构管理，如图 1-2 所示，用户视图下有系统视图，系统视图下还有接口视图、VLAN 视图等，不同视图对应功能分类，它们之间既有联系又有区别。当要使用某条命令配置某种功能时，需要先进入该命令所在的视图。用户想要了解某命令视图下支持哪些命令，可在该命令视图提示符下输入"？"，系统将自动罗列出该命令行视图下可以执行的所有命令。

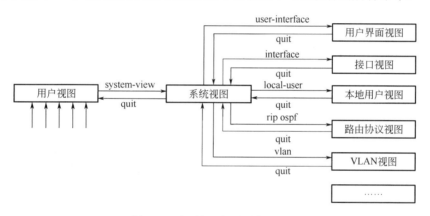

图 1-2　各种视图之间转换关系

用户视图：设备启动后，用户登录直接进入的缺省视图是用户视图，此时屏幕显示的提示符是：<设备名>。用户视图下可执行有限的操作命令，如查看设备启动后基本运行状态和统计信息，包括查看操作、调试操作、文件管理操作、设置系统时间、重启设备、FTP 和 Telnet 操作等。

系统视图：在用户视图下输入 system-view 命令进入系统视图。系统视图是全局操作界面，可以对设备运行参数进行配置，如修改系统时间、配置提示信息、配置快捷键等。在系统视图下输入不同的命令，可以进入相应的功能视图，完成各种功能的配置，比如：进入接口视图配置接口参数、创建 VLAN 并进入 VLAN 视图、进入用户界面视图配置登录用户的属性、创建本地用户并进入本地用户视图配置本地用户的密码和级别等。

用户界面视图：在系统视图下使用 user-interface 命令可进入用户界面视图。用户界面视图是配置、管理设备各用户界面属性的操作界面。在用户界面视图下网络管理员可以配置一系列参数，如用户登录时认证方式、用户登录后默认级别等。当用户使用该用户界面登录时，将受到这些参数的约束，从而达到统一管理各种用户会话连接的目的。

接口视图：在系统视图下使用 interface 命令可进入各种接口视图。接口视图是用来配置管理设备各种物理接口和逻辑接口的操作界面。在接口视图下可以配置以太网接口、同/异步接口、逻辑接口相关参数，如接口速率、IP 地址等。

本地用户视图：在系统视图下使用 local-user 可进入本地用户视图。它是用来配置管理本地用户属性的操作界面，包括创建本地用户账号，设置服务类型、用户登录密码。

初学者需注意用户视图、用户界面视图及本地用户视图三者区分。

路由协议视图：在系统视图下使用 RIP、OSPF 等命令可进入各种路由协议视图。路由协议视图是配置管理路由协议设置的操作界面，如创建 area ID 等。

VLAN：在系统视图下使用 VLAN ID 命令进入 VLAN 视图，在该视图下可配置三层虚接口地址。

可以使用帮助功能，在某一下视图下或者某一个命令的后面输入"？"，显示在该视图下或者在这个命令后面可以使用哪一些命令或参数。

在任意视图下可以使用 quit 命令退出此视图，切换到上一级视图。

在任意视图下使用<Ctrl+Z>或 return 命令直接退回用户视图。

3．命令行接口

H3C 网络设备的系统管理维护可通过基于 BootROM 的菜单操作、基于命令行接口的文本命令操作、基于 WEB 的图形界面操作三种方式来实现，文本命令操作是主要操作方式。

命令行接口 CLI（comand line interface，命令行接口）是用户与设备之间的文本类指令交互界面，用户输入文本类命令，通过输入回车键提交设备执行相关命令，从而对设备进行配置和管理，并可以通过查看输出信息确认配置结果，其命令格式如图 1-3 所示。对比图形用户界面 GUI（graphical user interface）使用鼠标点击相关选项进行设置，命令行接口形式可以一次输入含义更为丰富的指令，运用更灵活，功能更强大。

图 1-3　命令格式

4．超级终端和 Secure CRT 软件

超级终端是一款通用的串行交互软件，通过超级终端可以与网络设备系统实现交互，使超级终端成为网络设备系统运行的"显示器"。早期 Win XP 操作系统中自带超级终端程序组件，目前 Win7、Win8 常见操作系统默认情况下未安装超级终端程序，所以如需使用超级终端配置管理交换机、路由器等网络设备，安装 Win7、Win8 的计算机还需另外下载安装超级终端程序。

安装 Win7、Win8 的计算机还可以下载安装 Secure CRT 程序来配置管理交换机和路由器等网络设备。Secure CRT 是一款用于连接运行包括 Windows、UNIX 和 VMS 的终端仿真程序，是 Windows 登录 UNIX 或 Linux 服务器主机的软件。Secure CRT 的功能很全面而且使用方便，支持 Telnet，同时支持 SSH 和 Rlogin 协议。包括自动注册、对不同主机保持不同的特性、打印功能、颜色设置、可变屏幕尺寸、用户定义的键位图，能从命令行中运行或从浏览器中运行，还包括文本手稿、易于使用的工具条、用户的键位图编辑器、可定制的 ANSI 颜色等。Secure CRT 的 SSH 协议支持 DES、3DES 和 RC4 密码及 RSA 鉴别。

5．网络设备文件系统

网络设备文件系统的主要功能为管理存储设备。它把文件保存在相应存储设备中，并对存储设备中的文件、目录进行管理，包括创建并删除目录，显示当前的工作目录，显示指定目录下的文件或目录信息等目录管理操作；以及删除文件、恢复删除文件、彻底删除回收站中的文件、显示文件的内容、重新命名、复制文件、移动文件、执行批处理文件、显示指定文件的信息等文件管理操作。

五、实验过程

1．实验任务一：通过超级终端程序进行连接操作

① 通过专用配置线将 PC 串口与交换机或路由器 Console 口连起来。
② 启动超级终端程序并创建一个新连接,如图 1-4 所示，确定新连接名称。
③ 选择所用的通信串口，如图 1-5 所示。

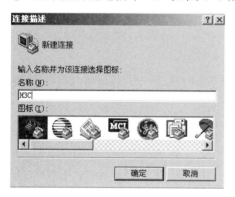

图 1-4　建立新连接

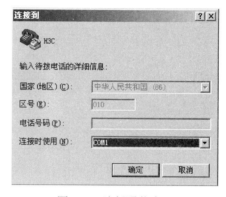

图 1-5　选择通信串口

④ 设置串口属性参数，如图 1-6 所示。设置波特率为 9600，数据位为 8，奇偶校验为无，停止位为 1，数据流控制为无。
⑤ 设置参数后，点击"确认"，进入网络设备的用户视图,如图 1-7 所示，实现网络设备登录。

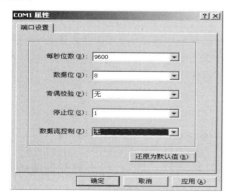

图 1-6　设置属性参数

图 1-7　用户视图

⑥ 登录网络设备后，重新启动网络设备可进入菜单操作界面。

```
<H3C>reboot                    // 重启命令。
```

界面将显示设备自检信息。

H3C MSR30-20 重启自检示例：

```
***************************************************************
*                                                               *
*            H3C MSR30-20  BootWare, Version 3.07               *
*                                                               *
***************************************************************
Copyright (c) 2004-2008 Hangzhou H3C Technologies Co., Ltd.

Compiled Date          : Jan  5 2009
CPU Type               : MPC8349E
CPU L1 Cache           : 32KB
CPU Clock Speed        : 533MHz
Memory Type            : DDR SDRAM
Memory Size            : 256MB
Memory Speed           : 266MHz
BootWare Size          : 4096KB
Flash Size             : 4MB
cfa0 Size              : 256MB
CPLD Version           : 2.0
PCB Version            : 3.0

BootWare Validating...
Press Ctrl+B to enter extended boot menu...
```

⑦ 若用户想修改成菜单操作启动，在 1s 内按下<Ctrl＋B>，进入 BOOT 菜单，显示如下：

```
BOOT  MENU
1. Download application file to flash    ←下载应用程序到 Flash 中
2. Select application file to boot       ←选择启动文件
3. Display all files in flash            ←显示 Flash 中的所有文件
4. Delete file from flash                ←删除 Flash 中的文件
5. Modify BootROM password               ←修改 BootROM 密码
6. Enter BootROM upgrade menu            ←进入 BootROM 升级菜单
7. Skip current configuration file       ←设置重启不运行当前配置文件
8. Set BootROM password recovery         ←恢复 BootROM 密码
9. Set switch startup mode               ←设置交换机启动模式
0. Reboot                                ←重新启动交换机
Enter your choice(0-9):
```

2. 实验任务二：通过 Secure CRT 软件进行连接操作

① 准备好交换机、配置线、COM 口转 USB 口转接器（图 1-8）。将配置线的 RJ45 口插入交换机的 Console 口，COM 口与转接器 COM 口连接，转接器 USB 口连接 PC，在 PC 上安装 Secure CRT 以及 COM 口转 USB 口转接器的驱动，实现交换机与 PC 连接。

② 打开 Secure CRT，新建连接，选择 Serial。可在 PC 的"设备管理器"中查看，COM 口转 USB 线的驱动安装的是哪个 COM 口，如图 1-9 所示。

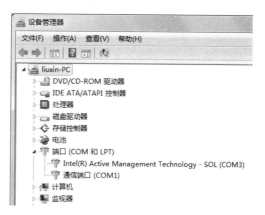

图 1-8　COM 口转 USB 转接器　　　　　图 1-9　从设备管理器查看 COM 口

③ 选择端口（选定 COM1，在"设备管理器"中查看确定），选择波特率为 9600，然后去掉流控的所有选项。进入 Secure CRT 界面，如图 1-10 所示。

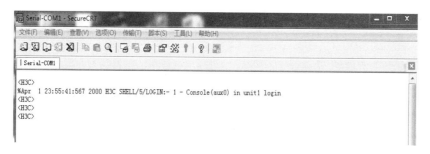

图 1-10　Secure CRT 界面

3. 实验任务三：实现视图之间的转换操作命令

（1）进入用户视图

用户登录设备后，直接进入用户视图。

```
<H3C>                          //屏幕显示的提示符是：<设备名>。
```

在用户视图中只能使用查看设备运行状态和统计信息的基本命令。包括查看操作、调试操作、文件管理操作、设置系统时间、重启设备、FTP 和 Telnet 操作等。

（2）进入系统视图

```
<H3C>system-view
[H3C]                          //在系统视图下配置系统全局通用参数。
```

如果有多个字母开头的命令，则按<Tab>键会逐个出现在屏幕上。

```
<H3C>sys                       //用<Tab>键在任何视图中都可以自动补全命令。
<H3C>system-view
```

（3）接口视图

[H3C]**interface Serial 6/0**
[H3C-Serial6/0]   //在接口视图下可以配置接口参数。

（4）路由协议视图

根据不同的路由协议进入该视图的命令也不同。

[H3C]**rip**   //使用 rip 协议则使用 RIP 命令。
[H3C-rip-1]
[H3C]**ospf**   //使用 ospf 协议则使用 OSPF 命令。
[H3C-ospf-1]

（5）用户界面视图

1）进入用户界面视图

命令格式：user-interface [type] first-number [last-number]

[H3C]**user-interface AUX 0**
[Router-ui-aux0]   //通过用户界面配置登录设备各个用户属性,统一管理各个用户。
[H3C]**user-interface VTY 0 4**   //同时进入 VTY 类型的 0~4 共 5 个用户界面。
[Router-ui-vty0-4]   //统一配置用户属性来管理各个用户。

2）显示用户界面的使用信息

命令格式：display users [all]

[H3C] **display users**
The user application information of the user interface(s):
Idx UI   Delay   Type Userlevel
+ 0 CON 0 00:00:00   3

  +   : Current operation user.
  F   : Current operation user work in async mode.

3）显示用户界面的物理属性和部分配置

命令格式：display user-interface [type number | number]

[H3C] **display user-interface AUX 0**
Idx Type   Tx/Rx   Modem Privi Auth Int
177 AUX 0   9600   -   0   P   -

  +   : Current user-interface is active.
  F   : Current user-interface is active and work in async mode.
  Idx   : Absolute index of user-interface.
  Type : Type and relative index of user-interface.
  Privi : The privilege of user-interface.
  Auth : The authentication mode of user-interface.
  Int   : The physical location of UIs.
  A   : Authentication use AAA.

```
L        : Authentication use local database.
N        : Current UI need not authentication.
P        : Authentication use current UI's password.
```

4）清除指定的用户界面

命令格式：free user-interface [type] number

4. 实验任务四：设备管理的基本操作命令

（1）修改设备名称

```
[H3C]sysname ?
TEXT  Host name (1 to 30 characters)        //设备名最长 30 个字符。
[H3C]sysname Router                          //将设备名称改为 Router。
[Router]
```

（2）修改系统时间

```
<Router>clock datetime ?
TIME  Specify the time (HH:MM:SS)           //时间格式为 HH:MM:SS。
<Router>clock datetime 14:50:30 ?
DATE  Specify the date from 2000 to 2035 (MM/DD/YYYY or YYYY/MM/DD)
//日期的格式为 MM/DD/YYYY 或者 YYYY/MM/DD。
<Router>clock datetime 11:50:30 2016/9/1
//修改日期为 2016/9/1  时间为 11:50:30。
<Router>display clock                        //显示系统时间。
11:51:00 UTC Thu 09/01/2016
```

（3）系统自动识别功能验证

在输入命令时为方便操作，有时仅输入前面的几个字符即可，但前提是这几个字符可以唯一表示一条命令。

```
<Router>dis clo                              //显示系统时间可以用。
15:14:43 UTC Fri 09/01/2016
```

（4）查看设备系统版本信息

```
<Router>display version
H3C COMWARE Platform Software
COMWARE Software, Version 5.20, Release 1719P02, Standard
Copyright (c) 2004-2009 Hangzhou H3C Tech. Co., Ltd. All rights reserved.
H3C MSR30-20 uptime is 0 week, 0 day, 0 hour, 9 minutes
Last reboot 2016/09/01 11:42:17
System returned to ROM By <Reboot> Command.

CPU type: FREESCALE MPC8349 533MHz
256M bytes DDR SDRAM Memory
4M bytes Flash Memory
Pcb              Version: 3.0
Logic            Version: 2.0
```

```
Basic     BootROM   Version:  3.07
Extended BootROM   Version:  3.07
[SLOT  0]CON            (Hardware)3.0,  (Driver)1.0,   (Cpld)2.0
[SLOT  0]AUX            (Hardware)3.0,  (Driver)1.0,   (Cpld)2.0
[SLOT  0]GE0/0          (Hardware)3.0,  (Driver)1.0,   (Cpld)2.0
[SLOT  0]GE0/1          (Hardware)3.0,  (Driver)1.0,   (Cpld)2.0
[SLOT  6]MIM-2SAE       (Hardware)2.0,  (Driver)1.0,   (Cpld)2.0
```

从显示信息中可以看出硬件平台为路由器 H3C MSR30-20，软件平台为 COMWARE 5.20 版本，BootROM 为 3.07 版本，CPU 频率为 533MHz，RAM 容量为 256M，闪存为 4M，提供了 SLOT 0 和 SLOT 6 两个槽，槽 SLOT 0 上具有 CON、AUX、GE0/0、GE0/1 四个端口，槽 SLOT 6 上安装 MIM-2SAE 模块。

（5）显示当前配置信息

```
[Router]display current-configuration
#
 version 5.20, Release 1719P02, Standard
#
 sysname Router
#
 domain default enable system
#
vlan 1
#
domain system
 access-limit disable
 state active
 idle-cut disable
 self-service-url disable
#
user-group system
#
local-user admin
 password cipher .]@USE=B,53Q=^Q`MAF4<1!!
 authorization-attribute level 3
 service-type telnet
#
interface Aux0
 async mode flow
 link-protocol ppp
#
interface Serial6/0
 link-protocol ppp
#
interface Serial6/1
 link-protocol ppp
#
interface NULL0
```

```
#
interface GigabitEthernet0/0
 port link-mode route
 ip address 192.168.1.1 255.255.255.0
#
interface GigabitEthernet0/1
 port link-mode route
#
user-interface con 0
user-interface aux 0
user-interface vty 0 4
#
return
```

5．实验任务五：文件、目录管理的基本操作命令

（1）保存配置命令

```
<Router>save                     //不指定存储文件名时，将默认存入 startup.cfg 文件。
The current configuration will be written to the device. Are you sure? [Y/N]:y
//选择 y，表示确定将当前配置文件写入存储介质中。
(To leave the existing filename unchanged, press the enter key):
Validating file. Please wait...
Now saving current configuration to the device.
Saving configuration flash:/startup.cfg. Please wait.....

[Router]save Router.cfg      //也可以通过保存时指定文件名来更改保存的文件名称。
The current configuration will be saved to flash:/Router.cfg. Continue? [Y/N]:y
Now saving current configuration to the device.
Saving configuration flash:/Router.cfg. Please wait....

<Router>display saved-configuration          //显示保存的配置文件内容。
#
 version 5.20, Release 1719P02, Standard
#
 sysname Router
#
 firewall enable
#
 nat address-group 1 202.100.138.100 202.100.138.105
#
 domain default enable system
#
acl number 2001
 rule 0 permit source 192.168.0.0 0.0.255.255
 rule 5 deny
```

```
#
acl number 3001
 rule 0 deny icmp source 192.168.10.0 0.0.0.255 destination 192.168.100.0 0.0.0.255
#
vlan 1
#
domain system
 access-limit disable
 state active
 idle-cut disable
 self-service-url disable
#
user-group system
#
local-user admin
 password cipher .]@USE=B,53Q=^Q`MAF4<1!!
 authorization-attribute level 3
 service-type telnet
#
interface Aux0
 async mode flow
 link-protocol ppp
#
interface Serial6/0
 link-protocol ppp
 nat outbound 2001 address-group 1
 ip address 202.100.138.1 255.255.255.0
#
interface Serial6/1
 link-protocol ppp
#
interface NULL0
#
interface GigabitEthernet0/0
 port link-mode route
 ip address 192.168.50.2 255.255.255.0
#
interface GigabitEthernet0/1
 port link-mode route
 firewall packet-filter 3001 outbound
 ip address 192.168.100.1 255.255.255.0
#
ospf 1 router-id 3.3.3.3
 area 0.0.0.0
  network 192.168.50.0 0.0.0.255
  network 192.168.100.0 0.0.0.255
```

```
#
user-interface con 0
user-interface aux 0
user-interface vty 0 4
#
return
```

如果设备没有保存配置文件，该命令不能显示任何配置信息。

（2）删除和清空配置文件

```
<Router>reset saved-configuration
The saved configuration file will be erased. Are you sure? [Y/N]:y
Configuration file in cfa0 is being cleared.
Please wait ...
 Configuration file is cleared.
```

通过这个命令可以清除配置。这时候使用 display saved-configuration 命令查看，设备显示没有配置文件。

```
<Router> display saved-configuration
#
 version 5.20, Release 1719P02, Standard
#
 sysname Router
#
 domain default enable system
#
vlan 1
#
domain system
 access-limit disable
 state active
 idle-cut disable
 self-service-url disable
#
user-group system
#
local-user admin
 password cipher .]@USE=B,53Q=^Q`MAF4<1!!
 authorization-attribute level 3
 service-type telnet
#
interface Aux0
  ---- More ----
```

配置信息还存在于当前运行设备的主存储器中，如使用 display current-configuration 命令查看，还可查看当前配置信息。

```
<Router> display current-configuration
#
```

```
 version 5.20, Release 1719P02, Standard
#
 sysname Router
#
 domain default enable system
#
vlan 1
#
domain system
 access-limit disable
 state active
 idle-cut disable
 self-service-url disable
#
user-group system
#
local-user admin
 password cipher .]@USE=B,53Q=^Q`MAF4<1!!
 authorization-attribute level 3
 service-type telnet
#
interface Aux0
```

<Router>**display current-configuration**
```
#
 version 5.20, Release 1719P02, Standard
#
 sysname Router
#
 domain default enable system
#
vlan 1
#
domain system
 access-limit disable
 state active
 idle-cut disable
 self-service-url disable
#
user-group system
#
local-user admin
 password cipher .]@USE=B,53Q=^Q`MAF4<1!!
 authorization-attribute level 3
 service-type telnet
#
interface Aux0
  ---- More ----
```

所以，还需要通过使用 reboot 命令重启设备，才能真正地清空当前配置。

（3）显示当前的工作目录

\<Router>**pwd**
cfa0:

表示当前路径是在 flash 存储器中。

（4）创建文件目录

命令格式：mkdir directory

\<Router>**mkdir newdirectory**

（5）显示当前路径的所有文件列表

\<Router>**dir**
Directory of cfa0:/

```
  0    -rw-    13340812   May 04 2008 12:19:02   main.bin
  1    drw-           -   Jul 24 2008 20:56:58   logfile
  2    -rw-    15858748   May 07 2009 18:57:44   msr30.bin
  3    -rw-       14209   Sep 01 2016 11:57:16   config.cwmp
  4    -rw-         759   Sep 01 2016 11:56:38   startup.cfg
  5    -rw-         759   Sep 01 2016 11:57:18   router.cfg
  6    drw-           -   Sep 01 2016 14:00:14   cfa0
  7    drw-           -   Sep 01 2016 14:00:50   newdirectory
```

254692 kB total （225852 kB free）
File system type of cfa0: FAT16

（6）改变当前工作路径

命令格式：cd directory

\<Router>**cd newdirectory**   //进入 newdirectory 子目录。

（7）删除文件目录

命令格式：rmdir directory

\<Router>**rmdir newdirectory**

（8）显示文本文件内容

\<Router>**more router.cfg**   //显示 router.cfg 文件中的内容。
#
 version 5.20, Release 1809P01, Standard
#
 sysname Router
#
 super password level 3 simple h3c
#

```
 domain default enable system
#
 telnet server enable
#
 dar p2p signature-file cfa0:/p2p_default.mtd
#
 port-security enable
#
vlan 1
#
domain system
 access-limit disable
 state active
 idle-cut disable
 self-service-url disable
#
user-group system
#
local-user admin
 password cipher .]@USE=B,53Q=^Q`MAF4<1!!
 authorization-attribute level 3
 service-type telnet
local-user Router
 password simple Router
 service-type telnet
#
interface Aux0
 async mode flow
 link-protocol ppp
#
interface Cellular0/0
 async mode protocol
 link-protocol ppp
#
interface Ethernet0/0
 port link-mode route
 ip address 192.168.1.2 255.255.255.0
#
interface Ethernet0/1
 port link-mode route
#
interface Serial1/0
 link-protocol ppp
#
interface Serial2/0
 link-protocol ppp
#
```

```
interface NULL0
#
 load xml-configuration
#
 load tr069-configuration
#
user-interface con 0
user-interface tty 13
user-interface aux 0
user-interface vty 0 4
 authentication-mode none
 user privilege level 3
 set authentication password simple h3c
#
return
```

(9)文件删除

```
<Router>dir
Directory of cfa0:/

   0     -rw-    13340812  May 04 2008 12:19:02   main.bin
   1     drw-           -  Jul 24 2008 20:56:58   logfile
   2     -rw-    15858748  May 07 2009 18:57:44   msr30.bin
   3     -rw-       14209  Sep 01 2016 11:57:16   config.cwmp
   4     -rw-         759  Sep 01 2016 11:56:38   startup.cfg
   5     -rw-         759  Sep 01 2016 11:57:18   router.cfg
   6     drw-           -  Sep 01 2016 14:00:14   cfa0
   7     drw-           -  Sep 01 2016 14:00:50   newdirectory
   8     drw-           -  Sep 01 2016 14:04:02   router

254692 kB total (225848 kB free)

File system type of cfa0: FAT16
<Router>delete router.cfg         //使用 delete 命令删除名为 router.cfg 的文件。
Delete flash:/router.cfg?[Y/N]:y
%Delete file flash:/router.cfg...Done.

<Router>dir                       //再次使用 dir 命令查看删除文件。
Directory of cfa0:/
   0     -rw-    13340812  May 04 2008 12:19:02   main.bin
   1     drw-           -  Jul 24 2008 20:56:58   logfile
   2     -rw-    15858748  May 07 2009 18:57:44   msr30.bin
   3     -rw-       14209  Sep 01 2016 11:57:16   config.cwmp
   4     -rw-         759  Sep 01 2016 11:56:38   startup.cfg
   5     drw-           -  Sep 01 2016 14:00:02   newirectory
254692 kB total (222932 kB free)
File system type of cfa0: FAT16
```

router.cfg 的文件没有显示，但是 flash 中的空间并没有变化。

```
<Router>dir /all                        //使用 dir /all 命令查看。
Directory of cfa0:/
   0    -rw-    13340812   May 04 2008 12:19:02   main.bin
   1    drw-           -   Jul 24 2008 20:56:58   logfile
   2    -rw-    15858748   May 07 2009 18:57:44   msr30.bin
   3    -rwh         336   Sep 01 2016 11:57:16   private-data.txt
   4    -rw-       14209   Sep 01 2016 11:57:16   config.cwmp
   5    -rw-         759   Sep 01 2016 11:56:38   startup.cfg
   6    drw-           -   Sep 01 2016 14:00:02   newdirectory
   7    drw-           -   Sep 01 2016 14:00:14   cfa0
   8    drw-           -   Sep 01 2016 14:00:50   newdirectory
   9    -rw-         759   Sep 01 2016 11:57:18   [router.cfg]
 254692 kB total  (222932 kB free)
File system type of cfa0: FAT16
```
router.cfg 文件仍然在 flash 中。

```
<Router>reset recycle-bin               //清空回收站。
Clear cfa0:/~/router.cfg ?[Y/N]:y
%Cleared file cfa0:/~/Router.cfg.
```
再次查看 flash
```
<Router>dir /all
Directory of cfa0:/

   0    -rw-    13340812   May 04 2008 12:19:02   main.bin
   1    drw-           -   Jul 24 2008 20:56:58   logfile
   2    -rw-    15858748   May 07 2009 18:57:44   msr30.bin
   3    -rwh         336   Sep 01 2016 11:57:16   private-data.txt
   4    -rw-       14209   Sep 01 2016 11:57:16   config.cwmp
   5    -rw-         759   Sep 01 2016 11:56:38   startup.cfg
   6    drw-           -   Sep 01 2016 14:00:14   cfa0
   7    drw-           -   Sep 01 2016 14:00:50   newdirectory
 254692 kB total (222940 kB free)
File system type of cfa0: FAT16
```
router.cfg 文件已经完全被删除了，而且 flash 空间也增加了。

六、实验思考

① 查阅资料，对比华为、思科、锐捷等公司网络设备系统管理维护的相关概念和操作命令的区别。

② 通过自检和信息查看操作，给出 H3C S3610 交换机详细配置信息，包括 CPU、存储器类型和容量、系统版本、接口类型。

③ H3C 网络设备系统主要文件类型有哪些？

实验二　网络设备登录及认证管理

一、实验目的

① 了解网络设备的本地配置与远程配置、本地认证与远程认证区别；
② 了解 H3C 系统提供的多种终端服务类型和用户类型；
③ 掌握网络设备通过 Console 口本地登录方法及认证方式的配置；
④ 掌握网络设备通过 Telnet 远程登录方法及认证方式的配置；
⑤ 熟悉浏览器或网管软件登录网络设备的配置管理方法。

二、实验内容

① 按图 2-1 连接实验设备，配置设备接口 IP 地址，实现设备与用户终端之间路由可达；
② 通过设备 Console 口本地登录，分别设置 password、scheme 登录认证方式的相关认证参数，保存配置文件后重启设备，并从 Console 口本地登录验证配置；
③ 通过 Console 口本地登录设备，启用 Telnet 服务，创建本地用户账号，设置密码、Telnet 服务类型、用户默认登录级别等认证参数，保存配置文件后重启设备，并从用户终端远程登录设备验证配置；
④ 通过 Console 口本地登录设备，启用 Web 服务，创建本地用户账号，设置密码、服务类型、用户默认登录级别等认证参数，利用浏览器或网管软件远程登录网络设备，实现图形界面配置管理。

三、实验设备及组网图

1. 实验设备

一台 H3C MSR30-20 路由器，一台 H3C S3610 交换机，一台计算机，超级终端或 Secure CRT 软件，一条专用 Console 配置线缆，一条交叉双绞线。

2. 组网图

网络设备登录及认证管理实验组网如图 2-1 所示。

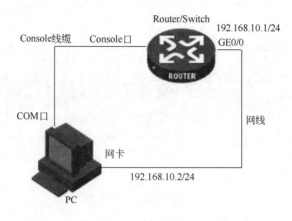

图 2-1 实验组网

四、实验相关知识

1．网络设备登录方法

H3C 系统提供多种终端服务，使用户可以登录设备进行配置管理。
① Console 口终端服务，通过专用线从 Console 口或 AUX 口进行本地登录配置；
② AUX 远程服务，通过 PSTN 挂接 Modem，从 AUX 口进行远程登录配置；
③ Telnet 服务，通过网络连接从设备网络接口进行远程登录配置；
④ SSH 服务，通过网络连接从设备网络接口进行远程登录配置；
⑤ 哑终端服务，通过计算机串口连接路由器异步口进行本地配置；
⑥ Web 服务，利用浏览器或网管软件登录网络设备进行远程登录配置。

通过 Console 口或 AUX 口进行本地登录配置又称为带外管理，通过网络接口（以太网接口、同/异步串口）使用 Telnet、SSH 等终端服务远程登录又称为带内管理。AUX 口除了可以如 CON 口一样进行本地配置外，还可以通过 PSTN 挂接 Modem 进行远程配置。

SSH（Secure Shell）是安全外壳的简称，用户在一个不能保证安全的网络环境远程登录网络设备时，SSH 特性可以提供安全保障和强大的认证功能，以保护网络设备不受诸如 IP 地址欺诈、明文密码截取等攻击。网络设备可作为 SSH 服务器端接受多个 SSH 客户的连接，也可作为 SSH 客户端与支持 SSH Server 的网络设备、Unix 主机等建立 SSH 连接。

哑终端工作方式是指当路由器的异步口（如同/异步串口、AUX 口、八异步口）在流方式下工作时，将主机或用户终端的串口与路由器异步口直连，可以进入路由器的命令行接口，对路由器进行配置管理。在哑终端基础上，可以建立其他应用，如执行 Telnet 命令登录其他设备。用户在 PC 上运行超级终端，可以与路由器任意一个异步口相连登录到路由器，对路由器进行配置管理。

2．用户分类与分级

在通常情况下，网络设备第一次启动时，系统没有设置用户名和登录口令。在不对登录用户进行验证的情况下，只要将计算机终端通过 Console 口与网络设备连接，任何用户都可以对网络设备进行配置管理。此时，一旦给设备主控板或接口板配置了 IP 地址，任何远端用户可以使用 Telnet 登录网络设备，远端用户还可能与网络设备建立 PPP 连接从而访问网络。这显然对网络设备和网络都是极不安全的。为此，需要为网络设备创建用户，设置用户口令，以此来对登录用户进行身份认证管理。

按用户所获得的应用服务，H3C 系统将用户划分为以下几种类型：
Terminal 用户，通过 Console 口或 AUX 口、异步口登录网络设备；
Telnet 用户，使用 Telnet 命令登录网络设备；
FTP 用户，与网络设备建立 FTP 连接进行文件传输；
PPP 用户，与网络设备建立 PPP 连接（如拨号、PPP oA 等），从而访问网络；
SSH 用户，与网络设备建立 SSH 连接，登录网络设备；
PAD 用户，与网络设备建立 PAD 连接，从而访问网络。

H3C COMWARE 系统采用分级保护方式将命令分为 4 个级别。访问级（Visit，0 级）：包括网络诊断工具命令和从本设备发出访问外部设备的命令，如 ping、tracert、telnet 等命令。监控级（Monitor，1 级）：用于系统维护、业务故障诊断等，包括 display、debugging 等命令，访问级和监控级都不能够进行配置文件的保存。系统级（System，2 级）：业务配置命令，包括各个层次网络协议的配置命令等。管理级（Manage，3 级）：最高级别，主要是关系到系统基本运行、系统支撑模块的命令。

为限制用户对系统的操作权限，COMWARE 系统对登录用户也进行了分级管理。用户的优先级也分为访问级、监控级、系统级、管理级 4 个级别，级别标识为 0～3。不同级别的用户只能使用包括他同级别及以下级别的命令。用户级别可通过命令进行切换，从低向高切换需要密码，从高向低切换不需密码。

3．用户登录认证方案

为保证网络设备安全，当用户登录路由器时，系统可对登录用户进行身份认证。H3C 网络设备对用户的认证设有 3 种方案：none、password 和 scheme。

none 认证方案：也就是不认证，即不需要进行用户名和密码认证，任何人都可以登录设备。但这种情况可能会带来安全隐患，为安全起见，网络设备实际应用中不建议采用 none 认证方案。

password 认证方案：表示用户登录设备时，只需要进行密码认证，不需用户名，只要密码认证成功，用户就能登录设备，可获得一定的安全性。password 认证模式及登录密码在用户界面视图下设置。

scheme 认证方案：表示用户登录设备时需要进行用户名和密码认证，用户名或密码错误，均会导致登录失败。scheme 认证模式在用户界面视图下设置，用户名和密码在本地用户视图下设置。

scheme 认证根据用户名和密码存放位置可分为本地认证和 AAA 服务器远程认证。本地认证由设备进行，用户提供的用户名和密码必须与登录设备上创建、存储的用户名和密码保持一致。AAA 服务器远程认证由 AAA 服务器进行，用户提供的用户名和密码必须与远程 AAA 服务器上创建、存储的用户名和口令一致。拨号用户常采用这种认证方案（本指导书不介绍基于 AAA 服务器远程认证）。

H3C 设备出厂时，不同的用户缺省的认证方式不同。Console 口登录模式缺省为 none 认证方式，便于用户初次登录。TTY、VTY 类型、AUX（远程配置）的用户界面缺省为 password 方式认证，其他类型用户界面缺省不进行终端认证。在配置 Telnet 用户和 Teminal 用户的认证方式时，通常设置为 scheme 方式中的本地认证方案。

不同的认证方式需要进行不同配置，常用配置命令如表 2-1 所示。

表 2-1　H3C 登录认证配置命令表

操　作	命　令
进入系统视图	system-view
进入 AUX 用户界面视图	user-interface aux *first-number* [*last-number*]
设置登录用户的认证方式为不认证	authentication-mode none
设置登录用户的认证方式为本地口令认证	authentication-mode password set authentication password { cipher | simple } *password*
设置登录用户的认证方式为通过认证方案认证	authentication-mode scheme
使能命令行授权功能	command-authorization
创建本地用户 （进入本地用户视图）	local-user *user-name*
设置本地用户认证口令	password { cipher | simple } *password*
设置本地用户的命令级别	authorization-attribute level *level*
设置本地用户的服务类型	service-type terminal

4．通过 Telnet 服务登录设备

Telnet 协议在 TCP/IP 协议族中属于应用层协议，通过网络提供远程登录和虚拟终端功能。在通常情况下，网络设备系统可提供 Telnet Server 服务，用户在主机上运行 Telnet 客户端程序登录可到网络设备上对设备进行配置管理，如图 2-2 所示。使用 Telnet 方式登录设备前，首先需要通过 Console 口登录设备，对认证方式、用户级别及公共属性进行相应的配置，才能保证通过 Telnet 方式正常登录设备。

图 2-2　网络设备提供 Telnet Server 服务

网络设备系统还可提供 Telnet Client 服务。如图 2-3 所示，在 PC 上通过终端仿真程序或 Telnet 程序登录连接的网络设备，再从该设备运行 Telnet Client 服务进一步登录与它连接的其他网络设备，并对其进行配置管理。

利用路由器这种重定向终端服务功能，用户通过 Telnet 客户端程序以特定的端口号登录路由器，可登录与路由器异步口相连的多个网络设备。典型的应用是实验室将路由器

的 8/16 异步口以直连方式外接多个设备，实现统一对多个设备进行远程配置和维护，如图 2-4 所示。

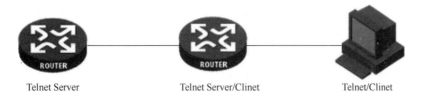

图 2-3　网络设备提供 Telnet Client 服务

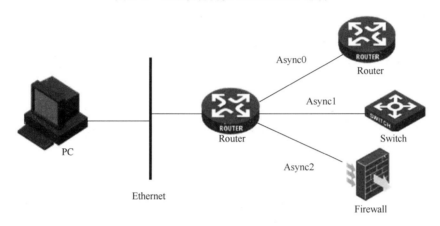

图 2-4　路由器重定向应用

五、实验过程

1. 实验任务一：搭建实验环境，初始化实验设备

按图 2-1 所示连接设备。登录设备后删除配置文件，再重启设备。建议每个实验任务前，清除配置文件并重启设备，完成实验设备初始化，避免原有设置参数影响实验结果。

```
<H3C>reset saved-configuration            //删除设备中的下次启动配置文件。
The saved configuration file will be erased. Are you sure? [Y/N]:y
Configuration file in cf is being cleared.
Please wait ...
........
 Configuration file in cf is cleared.     //清空配置，保持设备出厂设置。
<H3C>reboot                               //重启设备。
Start to check configuration with next startup configuration file, please wait.
........DONE!
This command will reboot the device. Current configuration may be lost in next
 startup if you continue. Continue? [Y/N]:y
```

```
#Apr  4 10:39:29:747 2010 Router_1 DEV/1/REBOOT:
 Reboot device by command.

%Apr  4 10:39:29:747 2010 Router_1 DEV/4/SYSTEM REBOOT:
 System is rebooting now.
 Now rebooting, please wait...
```

清空设备之后配置信息还保存在 RAM 中，需重启设备才能够完全清空配置。

<H3C>   //重启进入设备用户视图。

2. 实验任务二：配置 Console 口登录与认证方式

本实验任务选用 H3C MSR30-20 路由器。

在缺省情况下，用户可以直接通过 Console 口本地登录设备，登录时为 none 认证方式，不需要用户名和密码，登录用户级别为 3。用户成功登录设备后，如希望以后通过 Console 口登录设备时需要进行认证，可通过命令对设备登录模式进行修改。改变 Console 口登录方式的认证方式后，该认证方式的设置不会立即生效，需要退出命令行接口后重新登录，该设置才会生效。

（1）通过 password 认证方式登录配置

1）设置 password 认证方式和登录密码

[H3C]**sysname Router**   //对设备更名，本次实验对路由器进行配置。
[Router]**user-interface aux 0**   //进入 aux0 用户界面。
[Router-ui-aux0]**authentication-mode password**   //设置 password 认证方式。
[Router-ui-aux0]**set authentication password simple con_password**//设置密码。

2）设置 password 认证方式缺省登录用户级别和切换密码

[Router]**user-interface aux 0**
[Router-ui-aux0]**user privilege level 0**   //缺省登录用户级别为 0。
[Router-ui-aux0]**quit**
[Router]**super password level 1 simple x001**   //设置切换到用户级别 1 的密码。
[Router]**super password level 2 simple x002**
[Router]**super password level 3 simple x003**
[Router]**quit**

3）保存配置文件，重启网络设备验证配置

<Router> **save saved-configuration**
<Router>**reboot**   //重启设备验证配置。
……
password: //输入登录口令 con_password。
<Router>
<Router>super 1   //在用户视图下进行用户级别切换。
password: //输入用户级别切换 1 级密码 x001。
User privilege level is 1, and only those commands can be used

```
whose level is equal or less than this.
Privilege note: 0-VISIT, 1-MONITOR, 2-SYSTEM, 3-MANAGE
```

（2）用户 scheme 认证方式登录配置

1）设置 scheme 认证方式

```
[Router]user-interface aux 0
[Router-ui-aux0]undo set authentication password      //清除前面认证模式。
[Router-ui-aux0]undo user privilege level      //清除前面用户登录缺省级别。
[Router-ui-aux0]authentication-mode scheme      //设置 scheme 登录认证方式。
[Router-ui-aux0]quit
```

2）设置 scheme 认证方式用户账号、类型及缺省登录用户级别

```
[Router]local-user terminal_user      //创建新用户账号 terminal_user。
[Router-luser- terminal_user]service-type terminal
                                               //设置新用户账号服务类型为 terminal。
[Router-luser- terminal_user]password simple ter_user_password
[Router-luser- terminal_user]authorization-attribute level 2
                                               //设置登录缺省级别。
[Router-luser- terminal_user]display this  //查看本视图配置。
#
local-user terminal_user
 password simple ter_user_password
 service-type terminal
```

3）保存配置文件，重启设备验证配置

```
<Router>save
<Router>reboot
Users:                                         //输入用户账号 terminal_user。
password:                                      //输入登录口令 ter_user_password。
<Router>
```

3. 实验任务三：Telnet 远程登录网络设备

（1）按表 2-2 所示配置设备接口和主机 IP 地址。

表 2-2 IP 地址规划表

设备名称	接口	IP 地址	网关
Router	GE0/0	192.168.10.1/24	
PCA		192.168.10.2/24	

1）设置设备 E0/0 接口 IP 地址

```
[Router]interface GE0/0
[Router-GE0/0]ip address 192.168.10.1 24      //配置路由器接口 IP 地址。
```

24 表示网络号长度为 24，也可以用点分十进制表示：255.255.255.0。

2）设置主机的 IP 地址

在主机的本地连接中将 IP 地址设置为 192.168.10.2，子网掩码为 255.255.255.0。

（2）建立用户、设置密码、设置服务类型和用户级别

```
[Router]local-user telnet_user    //创建本地用户 telnet_user，进入本地用户视图。
New local user added.
[Router-luser-telnet_user]        //如本地用户已经存在，则直接进入本地用户视图。
[Router-luser-telnet_user]password simple tel_password   //为用户设置密码。
[Router-luser-telnet_user]service-type telnet    //设置服务类型为 telnet。
[Router-luser-telnet_user]authorization-attribute level 2
```

（3）配置 Telnet 用户的缺省认证方式

```
[Router]user-interface vty 0 4
```

H3C 系统支持 0~4 用户界面，支持 5 个 VTY 用户同时访问，可同时对 5 个 VTY 用户进行操作。

```
[Router-ui-vty0-4]authentication-mode scheme//设置本地认证授权方式为 scheme。
[Router]super password level 3 simple super3pass
                                                //设置切换到 level 3 的密码。
```

（4）设备开启 Telnet 服务

```
[Router]telnet server enable
% Start Telnet server
```

（5）用 Telnet 登录设备进行测试、验证

1）从 PC 上的 WIN7 打开"开始"菜单，选择"运行"，输入 cmd 命令，进入命令提示符窗口。

2）使用 Ping 命令检查主机与路由器的网络连通性。

```
G:\Documents and Settings\Administrator>ping 192.168.10.1
Pinging 192.168.10.1 with 32 bytes of data:
Reply from 192.168.10.1: bytes=32 time<1ms TTL=255
Reply from 192.168.10.1: bytes=32 time<1ms TTL=255
Reply from 192.168.10.1: bytes=32 time<1ms TTL=255
Reply from 192.168.10.1: bytes=32 time<1ms TTL=255

Ping statistics for 192.168.10.1:
    Packets: Sent = 4, Received = 4, Lost = 0 (0% loss),
Approximate round trip times in milli-seconds:
    Minimum = 0ms, Maximum = 0ms, Average = 0ms
```

表明 PC 与路由器的网络连通。

3）使用 Telnet 命令从主机登录设备。

```
G:\Documents and Settings\Administrator>telnet 192.168.10.1
```

```
Login authentication
Username:                          //输入 telnet 用户名 telnet_user。
Password:                          //输入 telnet 密码 tel_password。
<Router>?                          //查看当前视图可用命令。
User view     commands:
  cluster     Run cluster command
  display     Display current system information
  ping        Ping function
  quit        Exit from current command view
  rsh         Establish one RSH connection
  ssh2        Establish a secure shell client connection
  super       Set the current user priority level
  telnet      Establish one TELNET connection
  tracert     Trace route function
```

通过缺省用户名加密码的 scheme 方式登录路由器，命令级别为 0。

```
<Router>super 3                    //通过 super 命令切换到命令级别 3。
 Password:                         //输入用户切换到 level 3 的密码 super3pass。
User privilege level is 3, and only those commands can be used
whose level is equal or less than this.
Privilege note: 0-VISIT, 1-MONITOR, 2-SYSTEM, 3-MANAGE
<Router>?
User view commands:
  Archive        Specify archive settings
  backup         Backup next startup-configuration file to TFTP server
  boot-loader    Set boot loader
  BootROM        Update/read/backup/restore BootROM
  Cd             Change current directory
  clock          Specify the system clock
  cluster        Run cluster command
  copy           Copy from one file to another
  debugging      Enable system debugging functions
   ---- More ----
```

如果将 VTY 线路下的验证方式设置为 none，Telnet 登录时不需输入任何密码和用户名。

```
[Router-ui-vty0-4]authentication-mode none
[H3C-ui-vty0-4]user privilege level 3        //设置用户缺省命令级别为 3。
<Router>?
User view commands:
  archive           Specify archive settings
  backup            Backup next startup-configuration file to TFTP server
  boot-loader       Set boot loader
  BootROM           Update/read/backup/restore BootROM
  cd                Change current directory
  clock             Specify the system clock
  cluster           Run cluster command
  copy              Copy from one file to another
```

```
  debugging       Enable system debugging functions
  delete          Delete a file
  dialer          Dialer disconnect
  dir             List files on a file system
  display         Display current system information
 ---- More ----
```

直接设置 VTY 线路的认证方式为 password 时，将失去跟主机的连接。

```
Login password has not been set !        //发现显示无法和主机连接。
G:\Documents and Settings\Administrator>
```

在使用 password 认证方式时，需先在 VTY 线路下设置密码缺省命令级别。

[Router-ui-vty0-4]**authentication-mode password**
[Router-ui-vty0-4]**set authentication password simple password1**
[Router-ui-vty0-4]**user privilege level 3**
```
G:\Documents and Settings\Administrator>telnet 192.168.10.1
Password:                        //输入用户密码pass_word1。
<Router>?
User view commands:
  archive         Specify archive settings
  backup          Backup next startup-configuration file to TFTP server
  boot-loader     Set boot loader
  BootROM         Update/read/backup/restore BootROM
  cd              Change current directory
  clock           Specify the system clock
  cluster         Run cluster command
  copy            Copy from one file to another
  debugging       Enable system debugging functions
  delete          Delete a file
  dialer          Dialer disconnect
  dir             List files on a file system
  display         Display current system information
 ---- More ----
```

4．实验任务四：通过浏览器或网管软件登录网络设备

本实验任务选择 H3C S3610 交换机搭建 Web 网管远程配置管理环境。

H3C 网络设备系统提供内置的 Web 服务器，用户可以通过 PC 登录设备，使用 Web 界面直观地配置和维护设备。在缺省情况下，Web 服务器为启动状态，网络设备和 Web 网管终端都要进行相应的配置，才能保证通过 Web 网管正常登录设备。

（1）在以太网交换机上创建登录 Web 网管用户名和认证口令

通过 Console 口登录交换机，配置 Web 网管用户名为 admin，认证口令为 admin，用户级别为 3 级。

[H3C] **local-user admin**

```
[H3C-luser-admin] service-type telnet
[H3C-luser-admin] authorization-attribute level 3
[H3C-luser-admin] password simple admin
```

（2）通过 Console 口配置以太网交换机 VLAN 1 接口的 IP 地址

配置以太网交换机 VLAN 1 接口的 IP 地址为 192.168.10.2，子网掩码为 255.255.255.0。

```
<H3C> system-view
[H3C] interface vlan-interface 1
[H3C-VLAN-interface1] ip address 192.168.10.1 255.255.255.0
```

（3）启动或关闭 Web Server

操作命令如下。

```
<H3C>IP HTTP enable              //启动 Web Server。
<H3C>undo IP HTTP enable         //关闭 Web Server。
```

（4）用户通过 PC 与交换机相连

PC 的 IP 地址配置为 192.168.10.2（或确保 PC 终端和以太网交换机之间路由可达），在 Web 网管终端的浏览器地址栏内输入 http://192.168.10.1，浏览器会显示 Web 网管的登录页面，如图 2-5 所示。

图 2-5　Web 网管登录页面

（5）输入在交换机上添加的用户名和密码

点击"登录"按钮后即可登录，显示 Web 网管初始页面，如图 2-6 所示。

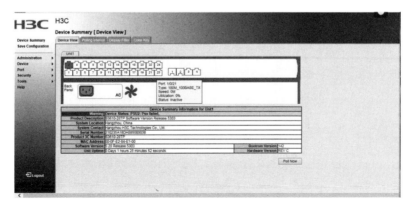

图 2-6　设备 Web 管理页面

（6）在设备上查看显示信息

在完成上述配置后，在任意视图下执行 display 命令可以显示 Web 用户的信息，通过 display web users 查看显示信息、验证配置的效果。

```
[H3C]display web users
UserID      Name      Language   Level         State    LinkCount   LoginTime   LastTime
ab7f0000    admin     English    Management    Enable   0           13:18:37    13:21:16
ab000100    admin     English    Management    Enable   0           13:25:54    13:25:59
```

五、实验思考

① 带内管理与带外管理的区别是什么？
② 密码设置命令中 simple 方式和 cipher 方式有什么区别？
③ 查阅设备相关操作手册，了解更多的设备管理登录方式。
④ 查阅设备相关操作手册，了解设备管理 AAA 认证方式。

实验三　网络设备系统软件升级

一、实验目的

① 理解 FTP 文件传输协议工作原理；
② 了解 H3C 网络设备系统软件升级方法；
③ 掌握使用 FTP 服务上传、下载文件基本操作。

二、实验内容

① 按图 3-1 连接实验设备，配置接口 IP 地址，实现设备与用户终端之间路由可达；
② 通过 Console 口本地进入网络设备 BootROM 菜单操作界面，通过 BootROM XModem 协议实现系统软件升级；
③ 以网络设备作为 FTP Server 端，以计算机为用户终端登录，升级 COMWARE 系统软件。

三、实验设备及组网图

1．实验设备

一台 H3C MSR30-20 路由器，一条专用 Console 配置线缆，一台计算机，一条双绞交叉线，超级终端或 Secure CRT 软件。

2．组网图

网络设备系统软件升级的试验组网如图 3-1 所示。

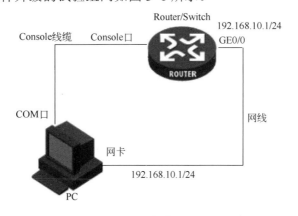

图 3-1　实验组网

四、实验相关知识

H3C 网络设备系统软件升级包括 BootROM 程序和 COMWARE 应用程序升级。H3C 网络设备使用的 COMWARE 软件将配套 BootROM 程序集成在内，因此升级 COMWARE 应用程序版本时系统会自动升级配套的 BootROM 程序，不需要单独进行 BootROM 升级。

在进行软件升级前，需要确认当前的 BootROM 版本及 COMWARE 应用程序版本，以便正确选择升级的程序文件，错误的应用程序版本可能导致网络设备不能正常启动。COMWARE 版本和 BootROM 程序版本配套关系可查阅网络设备厂商提供的说明书。H3C 网络设备系统升级具有多种方法。

1．通过 BootROM 菜单操作进行 H3C 软件升级

① 采用 BootROM XModem 方式的升级方法；
② 采用 BootROM FTP 方式的升级方法；
③ 采用 BootROM TFTP 方式的升级方法。

2．通过命令行操作方式进行 H3C 软件升级

① 网络设备作为 FTP Server 进行 COMWARE 版本升级的方法；
② 网络设备作为 FTP Client 进行 COMWARE 版本升级的方法；
③ 网络设备作为 TFTP Client 进行 COMWARE 版本升级的方法。

3．FTP 协议

FTP 协议在 TCP/IP 协议族中属于应用层协议，主要向用户提供与远程主机之间的文件传输，FTP 协议基于相应的文件系统实现。系统提供的 FTP 服务包括以下几种。

网络设备可作为 FTP 服务 Server 端升级 COMWARE 系统软件，用户可以在 PC 上运行 FTP 客户端程序登录路由器，访问管理路由器上的文件。

网络设备也作为 FTP 服务 Client 端。使用 FTP 升级 COMWARE 主体软件，用户通过终端仿真程序或 Telnet 程序建立与网络设备的连接后，使用 FTP 命令，再建立网络设备与远程 FTP Server 的连接，并访问管理远程主机上的文件。

在使用 FTP 服务时，需要 FTP 服务 Server 端启动 FTP 服务，FTP 客户端才能登录服务器，访问服务器上的文件。用户需通过验证和授权才能享受 FTP 服务器的服务功能，FTP 服务器的授权配置信息包含提供给 FTP 用户的工作目录路径的配置等，必须事先在网络设备上配置好用户类型和 FTP 工作目录。

4．TFTP 协议

TFTP（trivial file transfer protocol）是一种简单的文件传输协议。相对于文件传输协议 FTP，TFTP 不具有复杂的交互存取接口和认证控制，适用于客户机和服务器之间不需要复杂交互的环境。TFTP 协议一般在 UDP 的基础上实现。

TFTP 文件传输是由客户端发起的。当需要下载文件时，先由客户端向 TFTP 服务器

发送"读"请求包，然后从服务器接收数据包，并向服务器发送确认；当需要上传文件时，先由客户端向 TFTP 服务器发送"写"请求包，然后向服务器发送数据包，并接收服务器的确认。H3C 网络设备提供 TFTP 客户端和服务器端的功能。

5．XModem 协议

XModem 协议是一种文件传输协议，因其简单性和较好的性能而被广泛应用。

XModem 协议是通过串口传输文件，支持 128 字节和 1k 字节两种类型的数据包，并且支持一般校验和 CRC 校验两种方式，在出现数据包错误的情况下支持多次重传，一般为 10 次。

XModem 协议传输由接收程序和发送程序完成。先由接收程序发送协商字符，协商校验方式，协商通过之后，发送程序就开始发送数据包，接收程序接收到完整的一个数据包之后按照协商的方式对数据包进行校验，校验通过之后发送确认字符，然后发送程序继续发送下一包；如果校验失败，则发送否认字符，发送程序重传此数据包。

COMWARE 提供 XModem 接收程序功能，可以应用在 AUX 接口上，支持 128 字节的数据包和 CRC 校验。发送程序的功能自动包含在超级终端中。该协议可以用来升级 BootROM 程序、COMWARE 程序及配置文件。

五、实验过程

搭建实验环境，初始化实验设备。
按图 3-1 所示连接设备。登录设备后清除配置文件，再重启设备。

```
<H3C>reset saved-configuration
<H3C>reboot
……
<H3C>
```

1．实验任务一：通过 BootROM 菜单操作采用 BootROM XModem 协议升级软件系统

（1）路由器上电自检

用超级终端从 Console 口登录网络设备。

当路由器上电自检时，在出现"Press Ctrl-B to Enter Boot Menu..."的 3s 之内，按下 <Ctrl+B>，系统方可进入 BootROM 菜单。

（2）进入 BootROM 界面

进入 BootROM 界面后，系统提示"Please input BootROM password "，用户要输入 BootROM 口令（路由器出厂时缺省设置没有 BootROM 口令，直接回车即可）。若用户已经修改过 BootROM 口令，应该输入正确的口令，若三次口令验证未通过，则系统中止。

当用户输入正确的 BootROM 口令，系统将出现如下提示信息。

```
===================<EXTEND-BOOTROM MENU>===================

| <1> Boot From CF Card
```

```
| <2> Enter Serial SubMenu
| <3> Enter Ethernet SubMenu
| <4> File Control
| <5> Modify BootROM Password
| <6> Ignore System Configuration
| <7> Boot Rom Operation Menu
| <8> Clear Super Password
| <9> Device Operation
| <a> Reboot
=============================================================
Enter your choice(1-a):2
```

（3）选择 2 进入串口子菜单

```
 Enter Serial SubMenu
======================<SERIAL SUB-MENU>======================
|Note:the operating device is CF Card
| <1> Download Application Program To SDRAM And Run
| <2> Update Main Application File
| <3> Update Backup Application File
| <4> Update Secure Application File
| <5> Modify Serial Interface Parameter
| <0> Exit To Main Menu
=============================================================
Enter your choice(1-6):5
```

（4）选择 5 对串口的速率进行调整

选择 5，系统会提示修改串口波特率：

```
========================<BAUDRATE SET>=======================
|Note:  Change The HyperTerminal's Baudrate Accordingly,
|       Press 'Enter' to exit with things untouched.
|---------------------<Baudrate Avaliable>-------------------
| <1> 9600(Default)
| <2> 19200
| <3> 38400
| <4> 57600
| <5> 115200
| <0> Exit
=============================================================
Enter Your Choice(1-6):5
```

选择 1～5：以不同的波特率下载路由器 BootROM 软件。

（5）用户启用超级终端程序，使用 Xmodem 传送加载软件

如路由器的串口波特率已经选择为 115200bits/s，终端的波特率还为 9600bits/s，双方是无法通信的。所以根据上面提示，需修改终端设置的波特率为 115200bits/s，使其与所选的下载波特率一致，如图 3-2 所示。

单击"呼叫/呼叫"，重新连接。从终端窗口选择"传送/发送文件"，弹出对话框，如图 3-3 所示。

图 3-2 选择 Xmodem

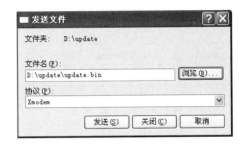

图 3-3 选择 Xmodem

单击"浏览"按钮，选择需要下载的应用程序文件，并将协议设置为 XModem，然后单击"发送"按钮，系统弹出对话框，如图 3-4 所示。

图 3-4 文件发送

下载完成后，终端界面出现如下信息，表明下载并升级成功：

```
Download successfully!
134432 bytes downloaded!
Updating Extend BTM
Updating Extended BootROM.
```

将配置终端的波特率从 115200bits/s 修改为 9600bits/s，重新启动路由器。

2．实验任务二：网络设备作为 FTP 服务端，通过命令行操作升级 COMWARE 系统软件

（1）清除配置文件，以出厂配置重启网络设备

```
<H3C>reset saved-configuration
<H3C>reboot
……
<H3C>
```

（2）按表 3-1 所示完成基本配置

表 3-1　IP 地址规划表

设备名称	接口	IP 地址	网关
Router	GE0/0	192.168.10.1/24	
PCA		192.168.10.2/24	

1）路由器接口 IP 地址配置

[Router]**interface GE0/0**
[Router-GE0/0]**ip address 192.168.10.1 24**

2）完成主机的配置

在主机的本地连接中将 IP 地址设置为 192.168.10.2，子网掩码为 255.255.255.0。

3）检查主机与路由器的连通性：在 Windows 操作系统下，单击"开始"，点击"运行"，输入"cmd"。

打开命令提示符窗口，输入 Ping 192.168.10.1 命令，检查网络连通性。

```
G:\Documents and Settings\Administrator>ping 192.168.10.1
Pinging 192.168.10.1 with 32 bytes of data:
Reply from 192.168.10.1: bytes=32 time<1ms TTL=255
Reply from 192.168.10.1: bytes=32 time<1ms TTL=255
Reply from 192.168.10.1: bytes=32 time<1ms TTL=255
Reply from 192.168.10.1: bytes=32 time<1ms TTL=255

Ping statistics for 192.168.10.1:
    Packets: Sent = 4, Received = 4, Lost = 0 (0% loss),
Approximate round trip times in milli-seconds:
    Minimum = 0ms, Maximum = 0ms, Average = 0ms
```

可以看到能够 Ping 通路由器，表示连通性没有问题。

（3）配置路由器

1）创建用户设置密码、用户级别和服务类型

[Router]**local-user ftp_user1**
[Router-luser- ftp_user1]**password simple ftp_password**
[Router-luser- ftp_user1]**service-type ftp**
[Router-luser- ftp_user1] **authorization-attribute level 3**

2）开启 ftp 服务功能

[Router]**ftp server enable**

（4）使用 FTP 登录路由器，进行上传下载文件操作

1）打开命令提示符

在 Windows 操作系统下，单击"开始"，点击"运行"，输入"cmd"，打开命令提示符。

2）登录到路由器

输入 FTP 命令及参数 IP 地址登录设备。

```
G:\Documents and Settings\Administrator>ftp 192.168.10.1
Connected to 192.168.10.1.
220 FTP service ready.
User (192.168.10.1:(none)): ftp_user1
331 Password required for Router.
Password: ftp_password
230 User logged in.
```

3）使用 ls 命令可以查看设备中存在的文件

```
ftp> ls
200 Port command okay.
150 Opening ASCII mode data connection for /*.
main.bin
logfile
msr30.bin
config.cwmp
newirectory
cfa0
newdirectory
router
226 Transfer complete.
```

ftp：收到 84B，用时 0.00s，速率为 84000.00kB/s。

可以看到设备中有四个文件，分别是 main.bin、newdirectory、default.diag、ma 文件。后缀名为.bin 的文件。

```
ftp> dir                    //使用 dir 命令可以查看文件详细属性。
200 Port command okay.
150 Opening ASCII mode data connection for /*.
-rwxrwxrwx  1  none  nogroup  13340812  May 04  2008 main.bin
drwxrwxrwx  1  none  nogroup         0  Mar 05  2009 newdirectory
-rwxrwxrwx  1  none  nogroup     44602  Oct 09  2009 default.diag
-rwxrwxrwx  1  none  nogroup       966  Sep 01  14:24 router.cfg
drwxrwxrwx  1  none  nogroup         0  Jan 19  13:19 ma
226 Transfer complete.
```

ftp：收到 524B，用时 0.01s，速率为 52400.00kB/s。

可以看到设备中文件的属性包括文件的权限、大小、创建时间等。

4）使用 get 命令下载配置文件 router.cfg

```
ftp> get router.cfg
200 Port command okay.
150 Opening ASCII mode data connection for /router.cfg.
```

226 Transfer complete.

ftp：收到 966B，用时 0.00s，速率为 966000.00kB/s。

默认下载到主机中当前用户的文件夹中，（如 C:\Documents and Settings\当前用户名。

查看主机中的文件，后面可以加上要下载到的指定盘符名和更改的文件名。例如，现在要将 Router.cfg 下载到本地电脑中的 D：盘，将名字改为 h3c.txt。

```
ftp> get router.cfg  d:/h3c.txt
200 Port command okay.
150 Opening ASCII mode data connection for /router.cfg.
226 Transfer complete.
```

ftp：收到 966B，用时 0.00s，速率为 966000.00kB/s。

ftp>

5）使用 put 命令上传文件

将 router.cfg 名称改为 config.cfg，使用 "put+文件名" 上传文件。

```
ftp> put config.cfg
200 Port command okay.
150 Opening ASCII mode data connection for /config.cfg.
226 Transfer complete.
```
ftp：发送 966B，用时 0.00s，速率为 966000.00kB/s。

```
<H3C>dir              //查看路由器中的文件。
Directory of cf:/

   0      -rw-   12629732  Sep 19 2008 14:24:16   main.bin
   1      drw-         -   Mar 05 2009 16:11:28   newdirectory
   2      -rw-      44602  Oct 09 2009 15:17:26   default.diag
   3      -rw-        966  Sep 02 2016 14:00:50   router.cfg
   4      -rw-        966  Sep 02 2016 14:01:58   config.cfg
254700 kB total (241448 kB free)
File system type of cf: FAT16
```

可以看到文件中多了 config.cfg 文件。

也可以在上传文件时更改文件的名称。方法是在之前的命令后加上要改的名称。如 put config.cfg wglab.txt 即可更改文件名为 config.cfg 。

```
ftp> put config.cfg  wglab.cfg
200 Port command okay.
150 Opening ASCII mode data connection for / wglab.cfg.
226 Transfer complete.
```

ftp：发送 966B，用时 0.00s，速率为 966000.00kB/s.

```
<H3C>dir              //查看路由器中的文件。
Directory of cf:/
   0      -rw-   12629732  Sep 19 2008 14:24:16   main.bin
   1      drw-         -   Mar 05 2009 16:11:28   newdirectory
```

```
2         -rw-        44602    Oct 09 2009 15:17:26    default.diag
3         -rw-          966    Sep 02 2016 14:00:50    Router.cfg
4         -rw-          966    Sep 02 2016 13:03:06    wglab.cfg
5         -rw-          966    Sep 02 2016 14:03:58    config.cfg
254700 kB total (241444 kB free)
File system type of cf: FAT16
```

文件中存在 wglab.cfg 文件，通过 ftp 服务实现了对文件的上传和下载。

六、实验思考

① 如何在 FTP 客户端中变换本地和远程的当前目录？
② 设备系统升级要注意什么事项？
③ BootROM 和 COMWARE 分别对应什么存储设备？

实验四 交换机端口设置与转发表维护

一、实验目的

① 熟悉 H3C 交换机命名规则、接口编号规则；
② 掌握交换机端口的基本配置；
③ 学习 MAC 地址转发表维护操作。

二、实验内容

① 识别 H3C S3610 交换机前、后面板接口类型、编号方式和指示灯含义；
② 按图 4-1 连接实验设备，完成连通性测试；
③ 学习端口速率、工作模式、流量控制等端口管理配置基本操作；
④ 学习 MAC 地址转发表的查看与维护操作。

三、实验设备及组网图

1. 实验设备

一台 H3C S3610 交换机，一条专用 Console 配置线缆，三台计算机，三条双绞线，超级终端或 Secure CRT 软件。

2. 组网图

交换机端口设置与转发表维护实验组网如图 4-1 所示。

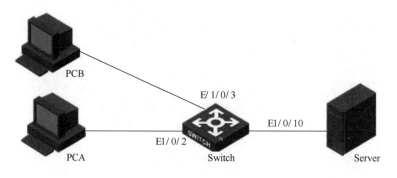

图 4-1 实验组网

四、实验相关知识

1．H3C 交换机命名规则

H3C 网络设备产品采用了一套字母或数字组合的命名规则。按规则命名的设备标识能清楚反映该产品类型、性能、基本配置、端口数量，以便用户和专业人员识别，如图 4-2 所示。

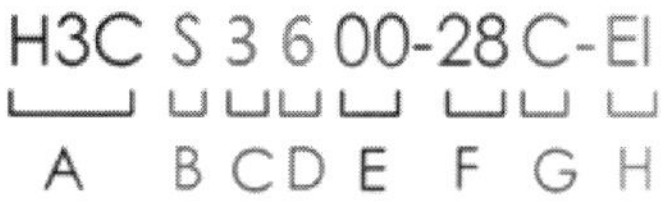

图 4-2 H3C 交换机命名规则

A 位表示产品品牌；
B 位表示产品系列：
  S——交换机   SR——业务路由器
C 位表示子产品系列：
  9——核心机箱式交换机   7——高端机箱式交换机
  5——全千兆盒式交换机   3——千兆以上交换机
D 位表示是否是路由交换机：
  $>=5$——路由交换机 <5——二层交换机
E 位表示低端：用于区别同一类的多个系列，高端是指业务槽位数
F 位表示可用端口数；
G 位表示上行接口类型：
  C——扩展插槽上行 P——千兆 SFP 光口上行 T——千兆电口上行
H 位表示业务特性：
  EI——增强型   SI——标准型
  PWR-EI——支持 PoE 的增强型 PWR-SI——支持 PoE 的标准型

2．H3C 交换机接口编号规则

H3C 系列交换机的接口名称由接口类型和接口编号两部分构成，接口编号采用 3 部分表示，如下所示：

Interface-type Unit ID/槽位编号/接口编号；

Interface-type：接口类型，取值可以为 Aux、Ethernet、GigabitEthernet、LoopBack、NULL 或 Vlan-interface。

Unit ID：单元号，是多台设备采用堆叠技术或智能弹性架构技术（intelligent resilient framework，IRF）时成员设备的编号。IRF 与堆叠技术相似，是将多台设备通过 IRF 物理端口连接在一起，通过软件虚拟化成一台"分布式设备"的技术。若使用堆叠或 IRF 技术时，Unit ID 取值范围为 1～8。若未使用堆叠或 IRF 技术时，如单台 H3C S3610 交换机运行该位取值为 1。

槽位编号：设备上的槽位号。一般设备上固有端口所在的槽位取值为 0；扩展接口模块卡 1 上端口所在的槽位取值为 1；扩展接口模块卡 2 上端口所在的槽位取值为 2，依此类推。如 H3C S3610 的百兆端口在 0 槽位上，千兆端口在 1 槽位上。

端口编号：某一槽位上的端口编号。

示例：Ethernet 1/0/1   GigabitEthernet 1/1/2 。

当用户配置相应以太网接口的相关参数时，必须使用 Interface 命令进入以太网接口视图。

命令格式：interface interface-type interface-number

可用 display brief interface 命令显示接口的简要配置信息，包括接口类型、连接状态、连接速率、双工属性、链路类型、缺省 VLAN ID、描述字符串。该命令的功能与 display interface 命令类似，只是显示的接口信息更加简要。

五、实验内容

1．搭建实验环境，初始化实验设备

按图 4-1 所示搭建实验环境，清除配置文件，以出厂配置重启网络设备。

```
<H3C>reset saved-configuration
<H3C>reboot
……
<H3C>
```

2．以太网端口的全双工/半双工设置

以太网交换机的端口存在全双工（full）、半双工（half）、自协商（auto）三种工作状态。duplex 命令用来设置以太网端口的双工属性。undo duplex 命令用来将端口的双工属性恢复为缺省的自协商状态。

命令格式：duplex { auto | full | half }

命令格式：undo duplex

auto：端口处于自协商状态。

full：端口处于全双工状态，端口在发送数据包的同时可以接收数据包。

half：端口处于半双工状态。端口同一时刻只能发送数据包或接收数据包。

```
<H3C>system-view
[H3C]interface Ethernet 1/0/1
[H3C-Ethernet1/0/1]duplex auto
[H3C-Ethernet1/0/1]undo duplex
[H3C-Ethernet1/0/1]display interface Ethernet 1/0/1   //查看一端口详细信息。
 Ethernet1/0/1 current state: UP
 IP Packet Frame Type: PKTFMT_ETHNT_2, Hardware Address: 000f-e284-e110
 Description: Ethernet1/0/1 Interface
 Loopback is not set
 Media type is twisted pair, Port hardware type is 100_BASE_TX
 100Mbps-speed mode, full-duplex mode
 Link speed type is autonegotiation, link duplex type is autonegotiation
 Flow-control is not enabled
 The Maximum Frame Length is 1552
 Broadcast MAX-ratio: 100%
```

```
 PVID: 1
 Mdi type: auto
 Port link-type: trunk
  VLAN passing  : 1 (default vlan)
  VLAN permitted: 1 (default vlan)
  Trunk port encapsulation: IEEE 802.1q
 Port priority: 0
 Last 300 seconds input:  0 packets/sec 0 bytes/sec
 Last 300 seconds output: 0 packets/sec 6 bytes/sec
 Input (total):  - packets, - bytes
        - broadcasts, - multicasts
 Input (normal): 32 packets, 3554 bytes
        13 broadcasts, 0 multicasts
 ---- More ----
```

3. 以太网端口的速率设置

以太网端口支持 10M、100M、1000M、自动协商等多种传输速率。当设置端口速率为自协商（auto）状态时，端口的速率由本端口和对端端口双方自动协商而定。对于 1000M 二层以太网端口，可以根据端口的速率自协商能力，指定自协商速率，让速率在指定范围内协商。speed 命令用来设置端口的速率。undo speed 命令用来恢复端口的速率为缺省值。在缺省情况下，端口速率处于自协商状态。

命令格式：speed { 10 | 100 | 1000 | auto }

命令格式：undo speed

10：指定端口速率为 10Mb/s。

100：指定端口速率为 100Mb/s。

1000：指定端口速率为 1000Mb/s（该参数仅适用于千兆端口）。

auto：指定端口的速率处于自协商状态。

```
[H3C-Ethernet1/0/1]speed 100
[H3C-Ethernet1/0/1]
%Apr 26 12:39:25:722 2000 H3C IFNET/4/LINK UPDOWN:
 Ethernet1/0/1: link status is DOWN
%Apr 26 12:39:27:419 2000 H3C IFNET/4/LINK UPDOWN:
 Ethernet1/0/1: link status is UP

[H3C-Ethernet1/0/1]display brief interface      //查看所有端口摘要信息。
The brief information of interface(s) under route mode:
Interface         Link   Protocol-link  Protocol type   Main IP
NULL0             UP     UP (spoofing)  NULL            --

The brief information of interface(s) under bridge mode:
Interface         Link   Speed    Duplex   Link-type  PVID
Eth1/0/1          DOWN   100M     auto     access     1
Eth1/0/2          DOWN   auto     auto     access     1
Eth1/0/3          DOWN   auto     auto     access     1
```

```
 Eth1/0/4          DOWN        auto      auto      access    1
 Eth1/0/5          DOWN        auto      auto      access    1
 Eth1/0/6          DOWN        auto      auto      access    1
 Eth1/0/7          DOWN        auto      auto      access    1
 Eth1/0/8          DOWN        auto      auto      access    1
 Eth1/0/9          DOWN        auto      auto      access    1
 Eth1/0/10         DOWN        auto      auto      access    1
 Eth1/0/11         DOWN        auto      auto      access    1
 Eth1/0/12         DOWN        auto      auto      access    1
 Eth1/0/13         DOWN        auto      auto      access    1
 Eth1/0/14         DOWN        auto      auto      access    1
 Eth1/0/15         DOWN        auto      auto      access    1
 Eth1/0/16         DOWN        auto      auto      access    1
 Eth1/0/17         DOWN        auto      auto      access    1
  ---- More ----
```

4．以太网端口的网线类型设置

用于连接以太网设备的双绞线主要有两类：直通线缆和交叉线缆。为了使以太网端口支持使用这两种线缆，设备实现了三种介质相关端口 MDI 模式：across、normal 和 auto。

物理以太网端口由 8 个引脚组成，在缺省状态下，每个引脚都有专门的作用，比如，使用引脚 1 和引脚 2 发送信号，引脚 3 和引脚 6 接收信号。通过设置 MDI 模式，可以改变引脚在通信中的角色。使用 normal 模式时，不改变引脚的角色，即使用引脚 1 和引脚 2 发送信号，使用引脚 3 和引脚 6 接收信号；如果使用 across 模式，会改变引脚的角色，将使用引脚 1 和引脚 2 接收信号，而使用引脚 3 和引脚 6 发送信号。只有将设备的发送引脚连接到对端的接收引脚后才能正常通信，所以 MDI 模式需要和两种线缆配合使用。

在通常情况下，建议用户使用 auto 模式，只有当设备不能获取网线类型参数时，才需要手工指定为 across 模式或 normal 模式。

当使用直通线缆时，两端设备的 MDI 模式配置不能相同。

当使用交叉线缆时，两端设备的 MDI 模式配置必须相同或者至少有一端设置为 auto 模式。

mdi 命令用来设置端口的 MDI 属性。undo mdi 命令用来恢复端口 MDI 属性的缺省值。

命令格式：mdi { across | auto | normal }

命令格式：undo mdi

across：设置端口的 MDI（介质相关接口）属性为只识别交叉网线。

auto：设置端口的 MDI 属性为自动识别网线类型。

normal：设置端口的 MDI 属性为只识别平行网线。

```
[H3C-Ethernet1/0/1]mdi across
[H3C-Ethernet1/0/1]display interface Ethernet 1/0/1
 Ethernet1/0/1 current state: UP
 IP Packet Frame Type: PKTFMT_ETHNT_2, Hardware Address: 000f-e284-e110
 Description: Ethernet1/0/1 Interface
 Loopback is not set
 Media type is twisted pair, Port hardware type is 100_BASE_TX
```

```
100Mbps-speed mode, full-duplex mode
Link speed type is force link, link duplex type is autonegotiation
Flow-control is not enabled
The Maximum Frame Length is 1552
Broadcast MAX-ratio: 100%
PVID: 1
Mdi type: across
Port link-type: trunk
 VLAN passing  : 1(default vlan)
 VLAN permitted: 1(default vlan)
 Trunk port encapsulation: IEEE 802.1q
Port priority: 0
Last 300 seconds input:  0 packets/sec 0 bytes/sec
Last 300 seconds output: 0 packets/sec 4 bytes/sec
Input (total):  - packets, - bytes
        - broadcasts, - multicasts
Input (normal): 34 packets, 3682 bytes
        15 broadcasts, 0 multicasts
```

5. 以太网端口广播风暴抑制比设置

交换机所有端口上的广播、组播或未知单播流量之和达到用户设置的值后，系统将丢弃超出广播、组播或未知单播流量限制的报文，从而使总体广播、组播或未知单播流量所占的比例降低到限定的范围，保证网络业务的正常运行。当单个端口上的广播、组播或未知单播流量超过用户设置的值后，系统将丢弃超出广播、组播或未知单播流量限制的报文，从而使该端口广播、组播或未知单播流量所占的比例降低到限定的范围，保证网络业务的正常运行。

用户可以通过 broadcast-suppression 命令限制以太网端口上允许通过的广播、组播或未知单播流量的大小。当在全局下进行配置时，设置的是整个系统允许通过的最大广播、组播或未知单播报文流量。当在端口下进行配置时，设置的是单个端口允许通过的最大广播、组播或未知单播报文流量。

命令格式：broadcast-suppression {ratio|pps max-pps }

命令格式：undo broadcast-suppression

ratio：指定以太网端口允许接收的最大广播流量的带宽百分比，取值范围为 1～100，缺省值为 100，步长为 1。百分比越小，允许接收的广播流量也越小。

max-pps：指定以太网端口每秒允许接收的最大广播包数量，单位为 pps。

在系统视图下，max-pps 的取值范围为 1～262143pps；

在以太网端口视图下，max-pps 的取值范围为 1～148810pps。

```
[H3C-Ethernet1/0/1]broadcast-suppression  30
[H3C-Ethernet1/0/1]display interface Ethernet 1/0/1
Ethernet1/0/1 current state: UP
IP Packet Frame Type: PKTFMT_ETHNT_2, Hardware Address: 000f-e284-e110
Description: Ethernet1/0/1 Interface
Loopback is not set
```

```
    Media type is twisted pair, Port hardware type is 100_BASE_TX
    100Mbps-speed mode, full-duplex mode
    Link speed type is force link, link duplex type is autonegotiation
    Flow-control is not enabled
    The Maximum Frame Length is 1552
    Broadcast MAX-ratio: 30%
    PVID: 1
    Mdi type: auto
    Port link-type: trunk
     VLAN passing  : 1(default vlan)
     VLAN permitted: 1(default vlan)
     Trunk port encapsulation: IEEE 802.1q
    Port priority: 0
    Last 300 seconds input:  0 packets/sec 0 bytes/sec
    Last 300 seconds output: 0 packets/sec 14 bytes/sec
    Input (total):  - packets, - bytes
          - broadcasts, - multicasts
    Input (normal): 34 packets, 3682 bytes
          15 broadcasts, 0 multicasts
    ---- More ----
```

6．开启以太网流量控制特性

flow-control 命令用来开启以太网端口流量控制特性。当开启端口流量控制后，设备具有发送和接收流量控制报文的能力，当本端发生拥塞时，设备会向对端发送流量控制报文；当本端接收到对端的流量控制报文后，会停止报文发送。undo flow-control 命令用来关闭以太网端口流量控制特性。

如配置 flow-control receive enable 命令，设备只具有接收流量控制报文的能力，但不具有发送流量控制报文的能力。当本端接收到对端的流量控制报文时，会停止向对端发送报文；当本端发生拥塞时，设备不能向对端发送流量控制报文。因此，如果要应对单向网络拥塞的情况，可以在一端配置 flow-control receive enable 命令，在对端配置 flow-control 命令；如果要求本端和对端网络拥塞都能处理，则两端都必须配置 flow-control 命令。

```
[H3C-Ethernet1/0/1]flow-control
[H3C-Ethernet1/0/1]display interface Ethernet 1/0/1
Ethernet1/0/1 current state: UP
IP Packet Frame Type: PKTFMT_ETHNT_2, Hardware Address: 000f-e284-e110
Description: Ethernet1/0/1 Interface
Loopback is not set
Media type is twisted pair, Port hardware type is 100_BASE_TX
100Mbps-speed mode, full-duplex mode
Link speed type is force link, link duplex type is autonegotiation
Flow-control is enabled
The Maximum Frame Length is 1552
Broadcast MAX-ratio: 100%
PVID: 1
Mdi type: auto
```

```
 Port link-type: trunk
  VLAN passing  : 1(default vlan)
  VLAN permitted: 1(default vlan)
  Trunk port encapsulation: IEEE 802.1q
 Port priority: 0
 Last 300 seconds input:  0 packets/sec 0 bytes/sec
 Last 300 seconds output:  0 packets/sec 7 bytes/sec
 Input(total): - packets, - bytes
       - broadcasts, - multicasts
 Input(normal): 34 packets, 3682 bytes
       15 broadcasts, 0 multicasts
 ---- More ----
```

7．关闭以太网端口

在某些特殊情况下，比如切换了端口的速率或双工模式等，端口相关配置不能立即生效，需要关闭和激活端口后才能生效。手工关闭端口，即便端口物理上是连通的，也不能转发报文。

shut down

undo shutdown

8．MAC 地址转发表管理

MAC 地址表记录了与本设备相连的设备的 MAC 地址，设备根据报文中的目的 MAC 地址查询 MAC 地址表，快速定位接口，从而减少广播。在一般情况下，MAC 地址表是设备通过源 MAC 地址学习过程而自动建立的。当通过源 MAC 地址学习自动建立 MAC 地址表时，设备无法区分合法用户和黑客用户的报文，带来了安全隐患。为了提高接口安全性，网络管理员可手工在 MAC 地址表中加入特定 MAC 地址表项，将用户设备与接口绑定，从而防止假冒身份的非法用户骗取数据。另外，如果需要丢弃指定源 MAC 地址或目的 MAC 地址的报文，可配置黑洞 MAC 地址表项。

配置动态/静态 MAC 地址表项命令格式如下：

mac-address {dynamic|static}mac-address interface interface-type interface-number vlan vlan-id

配置黑洞 MAC 地址表项命令格式如下：

mac-address blackhole mac-address vlan vlan-id

（1）设置交换机上动态 MAC 地址表项的老化时间为 500s。

[H3C] **mac-address timer aging 500**

（2）增加一个静态 MAC 地址和一个黑洞 MAC 地址（指出所属 vlan、端口、状态）。

[H3C] **mac-address static 00e0-fc35-dc71 interface Ethernet 1/0/2 vlan 1**
[H3C] **mac-address blackhole 0050-041D-A2B4 interface ethernet 1/0/3 vlan 1**

（3）查看设备 MAC 地址表信息。

[H3C]**display mac-address**

```
No Multicast Mac addresses found.
MAC ADDR          VLAN ID    STATE           PORT INDEX      AGING TIME(s)
000f-e2f3-5bc7    1          LEARNED         Ethernet1/0/24   AGING
0021-70fe-a294    1          LEARNED         Ethernet1/0/1    AGING
00d0-59cc-6053    1          LEARNED         Ethernet1/0/24   AGING

[H3C] display mac-address interface E1/0/2  //查看以太网接口 ethernet 1/0/2
MAC ADDR          VLAN ID    STATE           PORT INDEX      AGING TIME(s)
000f-e235-dc71    1          Config static   Ethernet 3/0/1   NOAGED
---  1 mac address(es) found  ---
```

六、实验思考

① 给出 H3C S3610 交换机某端口详细参数信息。
② MAC 地址转发表记录设置生存时间的作用是什么？
③ 交换机抑制广播帧的作用是什么？

实验五　链路聚合配置

一、实验目的

① 掌握交换机静态链路聚合的配置；
② 了解交换机动态链路聚合的配置；
③ 掌握验证链路聚合及信息分析的方法。

二、实验内容

① 按组网图连接实验设备，进行连通性测试；
② 实现以太网交换机静态链路聚合配置；
③ 实现以太网交换机动态链路聚合配置；
④ 验证链路聚合，分析查看信息。

三、实验设备及组网图

1．实验设备

两台 H3C 交换机，四台计算机，超级终端或 Secure CRT 软件，一条专用 Console 配置线缆，六条双绞线缆。

2．组网图

链路聚合配置试验组网如图 5-1 所示。

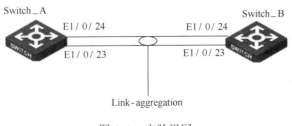

图 5-1　实验组网

四、实验相关知识

链路聚合是将多个物理以太网端口聚合在一起形成一个逻辑上的聚合组，使用链路聚合服务的上层实体把同一聚合组内的多条物理链路视为一条逻辑链路。链路聚合可以实现出/入负荷在聚合组中各个成员端口之间分担，以增加带宽。同时，同一聚合组的各个成员端口之间彼此动态备份，提高了连接可靠性。

根据成员端口上是否启用了 LACP 协议，可以将链路聚合分为静态聚合和动态聚合两种模式。

1．静态聚合

在静态聚合方式下，双方设备不需要启用聚合协议，双方不进行聚合组中成员端口状态的交互。如果一方设备不支持聚合协议或双方协议不兼容，则可以使用静态聚合。

静态聚合模式下的聚合组称为静态聚合组，当聚合组内有处于 up 状态的端口时，先比较端口的聚合优先级，优先级数值最小的端口作为参考端口；如果优先级相同，再按照端口的全双工/高速率—全双工/低速率—半双工/高速率—半双工/低速率的优先次序，选择优先次序最高，且第二类配置与对应聚合接口相同的端口作为该组的参考端口；如果优先次序也相同，则选择端口号最小的作为参考端口。

2．动态聚合

基于 IEEE802.3ad 标准的 LACP（Link Aggregation Control Protocol，链路聚合控制协议）是一种实现链路动态聚合的协议，运行该协议的设备之间通过互发 LACPDU（Link Aggregation Control Protocol Data Unit，链路聚合控制协议数据单元）来交互链路聚合的相关信息。动态聚合模式下的聚合组称为动态聚合组，动态聚合组内的选中端口以及处于 up 状态、与对应聚合接口的第二类配置相同的非选中均可以收发 LACPDU。

五、实验内容

1．实验任务一：交换机静态链路聚合配置

（1）按图 5-1 所示连接设备

登录设备后清除配置文件，再重启设备，避免原有设置参数影响实验结果。

```
<H3C>reset saved-configuration
<H3C>reboot
……
<H3C>
```

（2）静态聚合配置

先在系统视图下创建聚合端口，然后把物理端口加入聚合组中。
配置 Switch_A：

```
[H3C]sysname Switch_A
  [Switch_A]link-aggregation group 1 mode static/manual
  [Switch_A]interface Ethernet 1/0/23
    [Switch_A-Ethernet1/0/23]port link-aggregation group 1
  [Switch_A]interface Ethernet 1/0/24
    [Switch_A-Ethernet1/0/24]port link-aggregation group 1
    [Switch_A-Ethernet1/0/24]quit
```

配置 Switch_B：

```
[H3C]sysname Switch_B
  [Switch_B]link-aggregation group 1 mode static /manual
  [Switch_B]interface Ethernet 1/0/23
    [Switch_B-Ethernet1/0/23]port link-aggregation group 1
  [Switch_B]interface Ethernet 1/0/24
    [Switch_B-Ethernet1/0/24]port link-aggregation group 1
    [Switch_B-Ethernet1/0/24]quit
```

（3）查看聚合组配置信息

分别在 Switch_A 和 Switch_B 上查看所配置的聚合组信息。然后通过断开聚合组中的某条链路来观察网络连接是否中断。

```
[Switch_A]display link-aggregation summary
  Aggregation Group Type:S -- Static , M -- Manual
Loadsharing Type: Shar -- Loadsharing, NonS -- Non-Loadsharing
Actor ID: 0x8000, 000f-e284-dad0

 AL  AL   Partner ID              Select  Unselect  Share  Master
 ID  Type                         Ports   Ports     Type   Port
--------------------------------------------------------------------
 1   S    0x8000,000f-e284-e100    1       1         Shar   Ethernet1/0/24


[Switch_B]display link-aggregation summary
Aggregation Group Type:S -- Static , M -- Manual
Loadsharing Type: Shar -- Loadsharing, NonS -- Non-Loadsharing
Actor ID: 0x8000, 000f-e284-e100

 AL  AL   Partner ID              Select  Unselect  Share  Master
 ID  Type                         Ports   Ports     Type   Port
--------------------------------------------------------------------
 1   S    0x8000,000f-e284-dad0    1       1         Shar   Ethernet1/0/24
```

以上信息表明，交换机上有一个链路聚合端口，端口 ID 为 1，组中包含了 2 个端口，状态都为 Static（静态），并且工作在负责分担模式下。

（4）链路聚合组验证

在主机 PCA 上执行 Ping 命令，使 PCA 向 PCC 不间断发送 ICMP 报文，如下所示：

```
C:\Documents and Settings\Switch_pca>ping 192.168.10.3 -t
Pinging 192.168.10.3 with 32 bytes of data:
```

```
Reply from 192.168.10.3: bytes=32 time=1ms TTL=128
Reply from 192.168.10.3: bytes=32 time<1ms TTL=128
Reply from 192.168.10.3: bytes=32 time<1ms TTL=128
Reply from 192.168.10.3: bytes=32 time<1ms TTL=128
Reply from 192.168.10.3: bytes=32 time<1ms TTL=128
Reply from 192.168.10.3: bytes=32 time<1ms TTL=128
Reply from 192.168.10.3: bytes=32 time<1ms TTL=128
Reply from 192.168.10.3: bytes=32 time<1ms TTL=128
……
```

观察交换机面板上的端口 LED 显示灯，闪烁表示有数据流通过。将聚合组中 LED 显示灯闪烁的某个端口的线路断开，观察 PCA 上发送的 ICMP 报文是否丢失。

当一个端口不能转发数据流时，系统将数据流从另一个端口发送出去。在正常情况下，应该没有报文丢失。无报文丢失说明聚合组中的两个端口之间是互相备份的。

2. 实验任务二：交换机动态链路聚合配置

（1）清空 Switch_A 和 Switch_B 中的配置，确保 Switch_A 和 Switch_B 为初始状态

```
<H3C>reset saved-configuration
<H3C>reboot
……
<H3C>
```

（2）动态聚合配置

创建二层聚合端口 1，配置成动态聚合模式，并将端口加入聚合组一。

配置 Switch_A：

```
[Switch_A]link-aggregation group 1 description aaa
[Switch_A]interface Ethernet 1/0/23
[Switch_A-Ethernet1/0/23]lacp enable    //只有 S3610 S5510 交换机支持。
[Switch_A-Ethernet1/0/23]quit
[Switch_A]interface Ethernet 1/0/24
[Switch_A-Ethernet1/0/24]lacp enable
[Switch_A-Ethernet1/0/24]quit
```

配置 Switch_B：

```
[Switch_B] link-aggregation group 1 description aaa
[Switch_B]interface Ethernet 1/0/23
[Switch_B-Ethernet1/0/23]lacp enable
[Switch_B-Ethernet1/0/23]quit
[Switch_B]interface Ethernet 1/0/24
[Switch_B-Ethernet1/0/24]lacp enable
[Switch_B-Ethernet1/0/24]quit
```

（3）查看聚合配置信息

分别在 Switch_A 和 Switch_B 上查看所有配置的聚合组信息，如下所示：

```
[Switch_A]display link-aggregation summary
```

```
Aggregation Interface Type:
BAGG -- Bridge-Aggregation, RAGG -- Route-Aggregation
Aggregation Mode: S -- Static, D -- Dynamic
Loadsharing Type: Shar -- Loadsharing, NonS -- Non-Loadsharing
Actor System ID: 0x8000, 000f-e2f3-5db0

AGG         AGG       Partner ID              Select    Unselect    Share
Interface   Mode                              Ports     Ports       Type
--------------------------------------------------------------------------
BAGG1       D         0x8000, 000f-e2f3-5ba0  2         0           Shar

[Switch_B]display link-aggregation summary
Aggregation Interface Type:
BAGG -- Bridge-Aggregation, RAGG -- Route-Aggregation
Aggregation Mode: S -- Static, D -- Dynamic
Loadsharing Type: Shar -- Loadsharing, NonS -- Non-Loadsharing
Actor System ID: 0x8000, 000f-e2f3-5ba0

AGG         AGG       Partner ID              Select    Unselect    Share
Interface   Mode                              Ports     Ports       Type
--------------------------------------------------------------------------
BAGG1       D         0x8000, 000f-e2f3-5db0  2         0           Shar
```

以上信息表明，交换机上有一个链路聚合端口，端口 ID 为 1，组中包含了 2 个端口，状态都为 Dynamic（动态），并且工作在负责分担模式下。

（4）链路聚合组验证

通过这个实验掌握动态链路聚合的配置命令和查看方法。并且通过断开聚合组中的某条链路来观察网络连接是否中断。

在主机 PCA 上执行 Ping 命令，使 PCA 向 PCC 不间断发送 ICMP 报文，如下所示：

```
C:\Documents and Settings\Switch_pca>ping 192.168.10.3 -t
Pinging 192.168.10.3 with 32 bytes of data:
Reply from 192.168.10.3: bytes=32 time=1ms TTL=128
Reply from 192.168.10.3: bytes=32 time<1ms TTL=128
Reply from 192.168.10.3: bytes=32 time<1ms TTL=128
Reply from 192.168.10.3: bytes=32 time<1ms TTL=128
Reply from 192.168.10.3: bytes=32 time<1ms TTL=128
Reply from 192.168.10.3: bytes=32 time<1ms TTL=128
Reply from 192.168.10.3: bytes=32 time<1ms TTL=128
Reply from 192.168.10.3: bytes=32 time<1ms TTL=128
Reply from 192.168.10.3: bytes=32 time<1ms TTL=128
Reply from 192.168.10.3: bytes=32 time<1ms TTL=128
Reply from 192.168.10.3: bytes=32 time<1ms TTL=128
Reply from 192.168.10.3: bytes=32 time<1ms TTL=128
Reply from 192.168.10.3: bytes=32 time<1ms TTL=128
……
```

观察交换机面板上的端口 LED 显示灯，闪烁表示有数据流通过。将聚合组中 LED 显示灯闪烁的某个端口的线路断开，观察 PCA 上发送的 ICMP 报文是否丢失。

在正常情况下，应该没有报文丢失。无报文丢失说明聚合组中的两个端口之间是互相备份的。当一个端口不能转发数据流时，系统将数据流从另一个端口发送出去。

六、实验思考

① 链路聚合有什么作用？这种技术可以用于什么地方？
② 两台交换机直接用三根双绞线连接起来能正常工作吗？

实验六 STP 协议配置

一、实验目的

① 熟悉 STP 协议 BPDU 报文格式和转发机制；
② 了解交换机端口 STP 状态转换与指示灯之间的关系；
③ 掌握 STP 的基本配置操作。

二、实验内容

① 按图 6-1 所示连接实验设备，进行连通性测试；
② 启动生成树协议后查看交换机转发数据，分析相关数据信息。

三、实验设备及组网图

1．实验设备

两台 H3C 交换机，四台计算机，一条专用 Console 配置线缆，4 条双绞线缆，超级终端或 Secure CRT 软件。

2．组网图

STP 协议配置试验组网如图 6-1 所示。

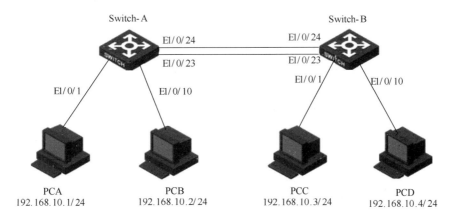

图 6-1　实验组网

四、实验相关知识

1. STP 的协议报文格式

启用 STP 协议的交换机使用 BPDU 特殊数据帧来传送设备的有关信息。它每隔一定的时间（缺省值为 2s）就发送和接受 BPDU 数据帧，其目的地址是多播 MAC 地址 01-80-C2-00-00-00，所有支持 STP 的网桥都会接收并处理收到的 BPDU 报文，用来检测生成树拓扑的状态变化，并通过生成树算法得到逻辑上没有环路生成树。BPDU 数据帧中包含了根信息、本网桥信息（网桥号）、路径花费和端口信息等，如图 6-2 所示。

| DMA | SMA | L/T | LLC Header | Payload |

值域	占用字节
协议 ID	2
协议版本	1
BPDU 类型	1
标志位	1
根桥 ID	8
根桥路径开销	4
指定桥 ID	8
指定端口 ID	2
MESSAGE AGE	2
MAX AGE	2
HELLO TOME	2
FORWARD DELAY	2

图 6-2 BPDU 协议报文格式

DMA：目的 MAC 地址，配置消息的目的地址是一个固定的桥的组播地址（0x0180c2000000）。

SMA：源 MAC 地址即发送该配置消息的桥 MAC 地址。

L/T：帧长。

LLC Header：配置消息固定的链路头。

Payload：BPDU 数据协议。

协议 ID：2 字节，当前保留没有被利用。

协议版本：1 字节，如果两个大小不一的协议版本数字比较,则数字越大的将被认为是最新定义的协议版本，当 STP 的版本为 IEEE 802.1d 时，值为 0。

BPDU 类型：1 字节，类型域仅仅服务于区分 BPDU 的类型；在不同类型 BPDU 之间没有任何关系，置 0x00 是配置 BPDU，置 0x08 是 TCN BPDU。在发生以下事件时，交换机发送 TCN。对于处于转发和监听状态的接口，过渡到 Block 状态（链路故障的情况），端口进入转发状态，并且网桥已经拥有指定端口，非根桥在它的指定端口收到 TCN。

标志位：1 字节，被用来表示拓扑的变化,当拓扑发生变化时，被置 1；反之，则置 0。

根桥 ID：8 字节，表示当前网络里的根桥，包括网桥优先级 2 字节，网桥的 MAC 地址为 6 字节。网桥优先级是用来衡量网桥在生成树算法中优先级的十进制数，取值范围是

0~65535，默认值是 32768，网桥 ID=网桥优先级+网桥 MAC 地址。

根路径开销：4 字节，根路径开销是两个网桥间的路径上所有链路的开销之和，也就是某个桥网到达根网桥的中间所有链路的路径开销之和。路径开销是 STP 协议用于选择链路的参考值。STP 协议通过计算路径开销，选择较为"强壮"的链路，阻塞多余的链路，将网络修剪成无环路的树型网络结构。

指定桥 ID：8 字节，由指定桥的优先级和 MAC 地址组成。

指定端口 ID：2 字节，由指定端口的优先级和该端口的全局编号组成，默认的端口优先级为 128。端口 ID=端口优先级+端口全局编号，端口优先级部分可配置。

Message Age：2 字节，配置消息在网络中传播的生存期。

Max AgeTime：2 字节，配置消息在设备中的最大生存期。在丢弃 BPDU 之前，网桥用来存储 BPDU 的时间，缺省为 20s。如果一个被阻塞的接口（非指定端口）在收到一个 BPDU 后，20s 内再没有收到 BPDU 了，则开始进入侦听状态。

Hello Time：2 字节，配置消息的发送周期；根网桥周期性发送配置 BPDU 的时间间隔，缺省为 2s。

Forward Delay Time：2 字节，端口状态迁移的延迟时间。转发延迟计时器，从侦听状态到学习状态，或者从学习状态转换到转发状态所需要的等待时间，缺省为 15s。

2．STP 端口状态转换

交换机完成启动后，如果交换机端口直接进入数据转发状态，而交换机此时并不了解所有拓扑信息，该端口可能会暂时造成数据环路。另外，链路故障会引发网络重新进行生成树的计算，生成树的结构将发生相应的变化。不过重新计算得到的新配置消息无法立刻传遍整个网络，如果新选出的根端口和指定端口立刻就开始数据转发的话，可能会造成暂时性的环路。为此，STP 采用了一种状态迁移的机制，引入了五种端口状态（图 6-3）及三个 BPDU 计时器：Forward Delay、Hello Time 和 Max Age。端口处于各种端口状态的时间长短取决于 BPDU 计时器，只有根桥的交换机可以通过生成树发送信息来调整计时器。

禁用状态（Disabled）：未启用或禁用 STP 功能的端口，端口不会参与生成树计算，也不会转发数据帧。

阻塞状态（Blocking）：端口处于只能接收状态，不参与数据帧的转发，但收听网络上的 BPDU 帧。该端口通过接收 BPDU 来判断根交换机的位置

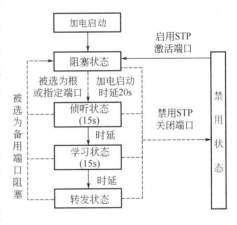

图 6-3　STP 端口状态转换

和根 ID，以及在 STP 拓扑收敛结束之后，各交换机端口应该处于什么状态，在默认情况下，端口会在这种状态下停留 20s。

侦听状态（Listening）：生成树此时已经根据本交换机所接收的 BPDU 而判断本端口应该参与数据帧的转发。于是交换机端口在接收 BPDU 的同时开始发送自己的 BPDU，并以此通告邻接的交换机，该端口会在活动拓扑中参与转发数据帧的工作。在默认情况下，该端口会在这种状态下停留 15s。

学习状态（Learning）：本端口不接收或转发数据帧，接收并发送 BPDU，学习 MAC 地址。在默认情况下，端口会在这种状态下停留 15s。

转发状态（Forwarding）：本端口已经成为活动拓扑的组成部分，它会转发数据帧，并同时收发 BPDU。

五、实验内容

1. 实验任务一：STP 基本实验

（1）搭建实验环境，初始化实验设备

按图 6-1 所示搭建实验环境，清除配置文件，以出厂配置重启网络设备。

```
<H3C>reset saved-configuration        //删除设备中的下次启动配置文件。
<H3C>reboot                           //重启设备。
……
<H3C>
```

（2）在系统视图下启用 stp

设置 Switch_A 的优先级为 4096，Switch_B 的优先级为默认值，并且配置连接 PC 的端口为边缘端口。

配置 Switch_A：

```
[H3C]sysname  Switch_A
[Switch_A]stp enable
[Switch_A]stp priority 4096              //配置时优先级必须是 4096 的倍数。
[Switch_A]interface Ethernet 1/0/1
[Switch_A-Ethernet1/0/1]stp edged-port enable    //用来连接主机端口。
```

配置 Switch_B：

```
[H3C]sysname  Switch_B
[Switch_B]stp enable
[Switch_B]interface Ethernet 1/0/1
[Switch_B-Ethernet1/0/1]stp edged-port enable
```

查看 stp 信息

分别在 Switch_A 和 Switch_B 上查看 stp 信息。

```
[Switch_A]display stp
-------[CIST Global Info][Mode MSTP]-------
CIST Bridge          :4096.000f-e284-e100
Bridge Times         :Hello 2s MaxAge 20s FwDly 15s MaxHop 20
CIST Root/ERPC       :4096.000f-e284-e100 / 0
CIST RegRoot/IRPC    :4096.000f-e284-e100 / 0
CIST RootPortId      :0.0
BPDU-Protection      :disabled
Bridge Config-
Digest-Snooping      :disabled
TC or TCN received   :4
Time since last TC   :0 days 0h:0m:27s
```

```
----[Port1（Ethernet1/0/1）][DOWN]----
Port Protocol           :enabled
Port Role               :CIST Disabled Port
Port Priority           :128
Port Cost（Legacy）     :Config=auto / Active=200000
Desg. Bridge/Port       :4096.000f-e284-e100 / 128.1
Port Edged              :Config=enabled / Active=enabled
Point-to-point          :Config=auto / Active=false
Transmit Limit          :10 packets/hello-time
Protection Type         :None
MST BPDU Format         :Config=auto / Active=legacy
[Switch_A]display stp brief
MSTID       Port                    Role   STP State    Protection
  0         Ethernet1/0/23          DESI   FORWARDING   NONE
  0         Ethernet1/0/24          DESI   FORWARDING   NONE
```

以上信息表明，Switch_A 是根桥，其上所有端口是指定端口（DESI），处于转发状态（Forwarding）。

```
[Switch_B]display stp brief
MSTID       Port                    Role   STP State    Protection
  0         Ethernet1/0/23          ROOT   FORWARDING   NONE
  0         Ethernet1/0/24          DESI   FORWARDING   NONE
```

以上信息表明，Switch_B 是非根桥，端口 Ethernet 1/0/23 是根端口，处于转发状态，端口 Ethernet 1/0/24 是备份端口（ALTE），处于阻塞状态。

2．选做实验任务：STP 综合实验

① 按图 6-4 所示搭建网络。清除设备配置文件，以出厂配置重启网络设备。
② 启动各交换机的 STP 功能，完成基本配置，实现生成树，并查看分析信息。

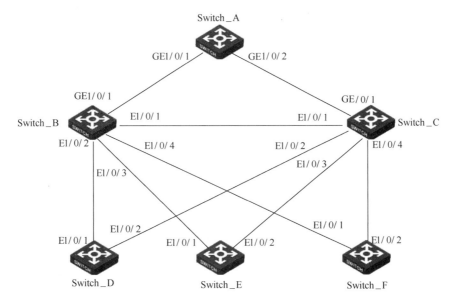

图 6-4　实验组网

六、实验思考

① 为什么交换机需要支持 STP 协议？
② 802.1D STP 有哪些缺点？
③ 三个 BPDU 计时器有什么作用？

实验七 VLAN 配置

一、实验目的

① 掌握基本 802.1Q 划分 VLAN 的实现原理；
② 掌握交换机 MAC 地址的学习过程；
③ 掌握 Access 链路端口、Trunk 链路端口和 Hybrid 链路端口的基本配置。

二、实验内容

① 按图 7-1 所示连接实验设备，完成连通性测试；
② 查看、分析 MAC 地址表中的表项，加深对 MAC 地址学习原理的理解；
③ 通过在单交换机划分 VLAN，验证交换机不同 VLAN 间的 PC 之间能否通信；
④ 通过跨交换机配置 Trunk 链路类型端口，验证跨交换机相同 VLAN、不同 VLAN 的 PC 之间的通信情况；
⑤ 通过配置 Hybrid 链路类型端口，实现不同 VLAN 之间的互通；
⑥ 查看分析 VLAN 数据。

三、实验设备及组网图

1．实验设备

两台 H3C S3610 交换机，一条专用 Console 配置线缆，四台计算机，超级终端或 Secure CRT 软件。

2．组网图

VLAN 配置实验组网如图 7-1 所示。

四、实验相关知识

1．IEEE 802.1Q 协议数据封装

如图 7-2 所示，传统的以太网数据帧在目的 MAC 地址和源 MAC 地址之后封装的是上层协议的类型字段。

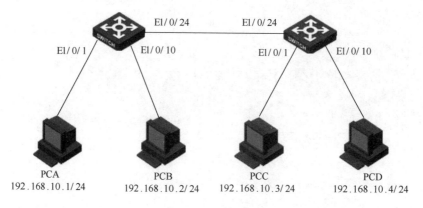

图 7-1 实验组网

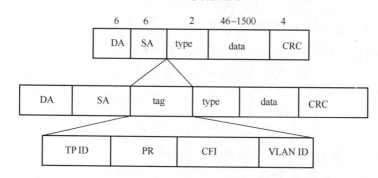

图 7-2 IEEE802.1Q 协议数据封装

其中 DA 表示目的 MAC 地址，SA 表示源 MAC 地址，Type 表示报文所属协议类型。

IEEE 802.1Q 协议规定在传统的以太网数据帧的目的 MAC 地址和源 MAC 地址之后封装 4 个字节的 VLAN Tag，用以标识 VLAN 的相关信息。

VLAN Tag 包含四个字段，分别是 TPID（Tag Protocol Identifier，标签协议标识符）、Priority、CFI（Canonical Format Indicator，标准格式指示位）和 VLAN ID。

TPID 用来标识本数据帧是带有 VLAN Tag 的数据，长度为 16 位，取值为 0x8100。

Priority 表示报文的 802.1Q 优先级，长度为 3 位。

CFI 字段标识 MAC 地址在不同的传输介质中是否以标准格式进行封装，长度为 1 位，取值为 0 表示 MAC 地址以标准格式进行封装，取值为 1 表示以非标准格式封装，缺省取值为 0。

VLAN ID 字段表示 VLAN 编号，长度为 12 位，范围从 0～4095，使用方法如表 7-1 所示。

表 7-1 VLAN 编号的有效范围表

VLAN	范 围	用 途
0 和 4095	保留	用户不能使用
1	正常范围	默认 VLAN，能使用，不能删除
2～1000	正常范围	用户创建、使用和删除
1001	正常范围	用户不能够创建、使用和删除

续表

VLAN	范围	用途
1002~1005	保留	提供 FDDI 和令牌环网使用
1006~1009	保留	
1010~1024	保留	
1025~4094	保留	有限使用

2．交换机端口工作模式

根据端口在转发数据帧时对 Tag 标签的不同处理方式，可将端口链路类型分为三种。

（1）Access 链路类型工作方式

Access 链路类型用于和不能识别 VLAN Tag 的终端设备相连，或者不需要区分不同 VLAN 成员时使用，所以 Access 链路类型端口发出去的数据帧不带 Tag 标签。一个 Access 端口只能属于一个 VLAN，可以通过手工设置指定 VLAN。

当 Access 端口接收到一个数据帧时，判断是否有 VLAN 信息，如果没有 VLAN 信息，则打上本端口的 PVID，并进行内部交换转发另一个端口；如果有 VLAN 信息，则直接丢弃。

Access 端口发送数据帧时，将数据帧的 VLAN 信息剥离，直接发送。

例如：交换机和普通的 PC 相连，PC 不能识别带 VLAN Tag 的数据帧，需要将交换机和 PC 相连端口的链路类型设置为 Access。

（2）Trunk 链路类型工作方式

Trunk 链路类型通常用于网络传输设备之间的互联。一个 Trunk 口在缺省情况下属于本交换机所有 VLAN，它能够转发所有 VLAN 的帧，但是可以通过设置许可 VLAN 列表加以限制。

Trunk 端口接收到一个数据帧，判断是否有 VLAN 信息，如果没有，则打上本端口的 PVID，并内部交换转发另一个端口；如果有其他 VLAN 标签，判断该 Trunk 端口是否允许该 VLAN 的数据进入，如果可以，则带 VLAN 标签内部交换转发另一个端口，否则丢弃。

Trunk 端口发送数据帧时，将要发送数据帧的 VLAN 与端口 PVID 比较，如果两者相等，则剥离 VLAN 信息，再发送；如果不相等，判断该 Trunk 端口是否允许该 VLAN 的数据通过，如允许则直接发送，否则丢弃。

（3）Hybrid 链路类型工作方式

Hybrid 链路类型可以用于交换机之间的连接，也可以用于连接用户的计算机。Hybrid 类型的端口可以允许多个 VLAN 通过，可以接收和发送多个 VLAN 的数据帧，端口发送的数据帧可根据需要设置某些 VLAN 内的数据帧带 Tag，某些 VLAN 内的数据帧不带 Tag。

Hybrid 端口接收到一个数据帧时，判断是否有 VLAN 信息，如果没有，则打上端口的 PVID，并进行交换转发；如果有，则判断该 Hybrid 端口是否允许该 VLAN 的数据进入，如果可以则转发，否则丢弃，此时端口上的 Untag 配置是不用考虑的，Untag 配置只对发送数据帧起作用。

Hybrid 端口发送数据帧时，判断该 VLAN 在本端口的属性；如果是 Untag，则剥离 VLAN 信息，再发送；如果是 Tag，则直接发送。

五、实验内容

1. 实验任务一：单交换机上 VLAN 划分的实现

（1）搭建实验环境，初始化实验设备

按图 7-1 所示搭建实验环境，清除配置文件，以出厂配置重启网络设备。

```
<H3C>reset saved-configuration
<H3C>reboot
......
<H3C>
```

（2）按表 7-2 所示配置设备名和主机 IP 地址

表 7-2 IP 地址规划及 VLAN 划分

设备名称	接口	IP 地址	VLAN
Switch_A	Ethernet 1/0/1 to Ethernet 1/0/9		VLAN 10
	Ethernet 1/0/10 to Ethernet 1/0/20		VLAN 20
PCA	Ethernet 1/0/1	192.168.10.1	VLAN 10
PCB	Ethernet 1/0/10	192.168.10.2	VLAN 20

1）连通性测试，在主机 PCA 上 Ping 主机 PCB

```
C:\Documents and Settings\Switch_pca>ping 192.168.10.2
Pinging 192.168.10.3 with 32 bytes of data:
Reply from 192.168.10.3: bytes=32 time<1ms TTL=128
Reply from 192.168.10.3: bytes=32 time<1ms TTL=128
Reply from 192.168.10.3: bytes=32 time<1ms TTL=128
Reply from 192.168.10.3: bytes=32 time<1ms TTL=128
......
```

2）交换机缺省 VLAN 信息

```
[Switch_A]display vlan all                //查看交换机上的 VLAN 信息。
VLAN ID: 1
VLAN Type: static
Route Interface: not configured
Description: VLAN 0001
 Tagged   Ports: none
 Untagged Ports:
```

```
            Ethernet1/0/1          Ethernet1/0/2          Ethernet1/0/3
            Ethernet1/0/4          Ethernet1/0/5          Ethernet1/0/6
            Ethernet1/0/7          Ethernet1/0/8          Ethernet1/0/9
            Ethernet1/0/10         Ethernet1/0/11         Ethernet1/0/12
            Ethernet1/0/13         Ethernet1/0/14         Ethernet1/0/15
            Ethernet1/0/16         Ethernet1/0/17         Ethernet1/0/18
            Ethernet1/0/19         Ethernet1/0/20         Ethernet1/0/21
            Ethernet1/0/22         Ethernet1/0/23         Ethernet1/0/24
            GigabitEthernet1/1/1   GigabitEthernet1/1/2   GigabitEthernet1/1/3
            GigabitEthernet1/1/4
[Switch_A]display brief interface        //查看交换机上的接口信息。
The brief information of interface(s) under route mode:
Interface           Link      Protocol-link  Protocol type    Main IP
NULL0               UP        UP(spoofing)   NULL             --

The brief information of interface(s) under bridge mode:
Interface           Link      Speed      Duplex      Link-type    PVID
Eth1/0/1            UP        100M(a)    full(a)     access       1
Eth1/0/2            DOWN      auto       auto        access       1
Eth1/0/3            DOWN      auto       auto        access       1
Eth1/0/4            DOWN      auto       auto        access       1
Eth1/0/5            DOWN      auto       auto        access       1
Eth1/0/6            DOWN      auto       auto        access       1
Eth1/0/7            DOWN      auto       auto        access       1
Eth1/0/8            DOWN      auto       auto        access       1
Eth1/0/9            DOWN      auto       auto        access       1
Eth1/0/10           DOWN      auto       auto        access       1
---- More ----
```

从上面信息可知，交换机上的缺省 VLAN 是 VLAN1，即端口的 PVID 是 1，以出厂配置所有的端口处于 VLAN1 中，且端口类型都是 Access 链路类型。

（3）按表 7-2 所示划分 VLAN

配置 Switch_A：

```
[Switch_A]vlan 10        //创建 VLAN 10 并进入其 VLAN 视图。
[Switch_A-vlan10]port Ethernet 1/0/1 to Ethernet 1/0/9
                         //将交换机的 1 至 9 端口划入 vlan 10 中。
[Switch_A-vlan10]vlan 20
[Switch_A-vlan20]port Ethernet 1/0/10 to Ethernet 1/0/19
[Switch_A-vlan20]quit
```

（4）查看、分析交换机的 VLAN 信息

```
[Switch_A]display vlan all
VLAN ID: 1
VLAN Type: static
Route Interface: not configured
Description: VLAN 0001
```

```
    Tagged    Ports: none
  Untagged Ports:
    Ethernet1/0/21          Ethernet1/0/22          Ethernet1/0/23
    Ethernet1/0/24
    GigabitEthernet1/1/1    GigabitEthernet1/1/2    GigabitEthernet1/1/3
    GigabitEthernet1/1/4

  VLAN ID: 10
  VLAN Type: static
  Route Interface: not configured
  Description: VLAN 0010
    Tagged    Ports: none
  Untagged Ports:
    Ethernet1/0/1           Ethernet1/0/2           Ethernet1/0/3
    Ethernet1/0/4           Ethernet1/0/5           Ethernet1/0/6
    Ethernet1/0/7           Ethernet1/0/8           Ethernet1/0/9

  VLAN ID: 20
  VLAN Type: static
  Route Interface: not configured
  Description: VLAN 0020
    Tagged    Ports: none
  Untagged Ports:
    Ethernet1/0/10          Ethernet1/0/11          Ethernet1/0/12
    Ethernet1/0/13          Ethernet1/0/14          Ethernet1/0/15
    Ethernet1/0/16          Ethernet1/0/17          Ethernet1/0/18
    Ethernet1/0/19
```

以上信息表明，端口 Ethernet1/0/20～Ethernet1/0/24 及端口 GigabitEthernet1/1/1～GigabitEthernet1/1/4 仍然属于缺省 VLAN 1 中；端口 Ethernet1/0/1～Ethernet1/0/9 被成功划分到 VLAN 10 当中，并且都在 Untagged 列表中；端口 Ethernet1/0/10～Ethernet1/0/19 被成功划分到 VLAN 20 当中，并且都在 Untagged 列表中。

（5）连通性测试

PCA 属于 VLAN 10，PCB 属于 VLAN 20，在主机 PCA 上 Ping 主机 PCB，结果如下所示：

```
C:\Documents and Settings\Switch_pca>ping 192.168.10.2
Pinging 192.168.10.2 with 32 bytes of data:

Request timed out.
Request timed out.
Request timed out.
Request timed out.
……
```

通过上面测试结果可知，不同 VLAN 间的主机不能互通。

2. 实验任务二：通过 Trunk 链路类型端口实现跨交换机 VLAN 划分

（1）按表 7-3 所示配置设备名、主机 IP 地址并划分 VLAN

表 7-3　IP 地址规划及 VLAN 划分

设备名称	接口	IP 地址	VLAN
Switch_A	Ethernet 1/0/1 to Ethernet 1/0/9		VLAN 10
	Ethernet 1/0/10 to Ethernet 1/0/20		VLAN 20
Switch_B	Ethernet 1/0/1 to Ethernet 1/0/9		VLAN 10
	Ethernet 1/0/10 to Ethernet 1/0/20		VLAN 20
PCA	Ethernet 1/0/1	192.168.10.1	VLAN 10
PCB	Ethernet 1/0/10	192.168.10.2	VLAN 20
PCC	Ethernet 1/0/1	192.168.10.3	VLAN 10
PCD	Ethernet 1/0/10	192.168.10.4	VLAN 20

（2）配置 Trunk 链路端口

在交换机 Switch_A、Switch_B 上配置端口 Ethernet 1/0/24 为 Trunk 链路类型端口。
配置 Switch_A：

```
[Switch_A] interface Ethernet 1/0/24
[Switch_A-Ethernet1/0/24]port link-type trunk
[Switch_A-Ethernet1/0/24]port trunk permit vlan all
```

配置 Switch_B：

```
[Switch_B]interface Ethernet 1/0/24
[Switch_B-Ethernet1/0/24]port link-type trunk
[Switch_B-Ethernet1/0/24]port trunk permit vlan all
```

查看所有 VLAN 信息，如下所示：

```
[Switch_A]display vlan all
VLAN ID: 1
VLAN Type: static
Route Interface: not configured
Description: VLAN 0001
Tagged   Ports: none
Untagged Ports:
  Ethernet1/0/21         Ethernet1/0/22          Ethernet1/0/23
  Ethernet1/0/24
  GigabitEthernet1/1/1   GigabitEthernet1/1/2    GigabitEthernet1/1/3
  GigabitEthernet1/1/4

VLAN ID: 10
VLAN Type: static
Route Interface: not configured
```

```
 Description: VLAN 0010
 Tagged   Ports:
   Ethernet1/0/24
 Untagged Ports:
   Ethernet1/0/1         Ethernet1/0/2         Ethernet1/0/3
   Ethernet1/0/4         Ethernet1/0/5         Ethernet1/0/6
   Ethernet1/0/7         Ethernet1/0/8         Ethernet1/0/9

 VLAN ID: 20
 VLAN Type: static
 Route Interface: not configured
 Description: VLAN 0020
 Tagged   Ports:
   Ethernet1/0/24
 Untagged Ports:
   Ethernet1/0/10        Ethernet1/0/11        Ethernet1/0/12
   Ethernet1/0/13        Ethernet1/0/14        Ethernet1/0/15
   Ethernet1/0/16        Ethernet1/0/17        Ethernet1/0/18
   Ethernet1/0/19        Ethernet1/0/20
[Switch_A]display interface Ethernet 1/0/24
  Ethernet1/0/24 current state: UP
  IP Packet Frame Type: PKTFMT_ETHNT_2, Hardware Address: 000f-e284-e127
  Description: Ethernet1/0/24 Interface
  Loopback is not set
  Media type is twisted pair, Port hardware type is 100_BASE_TX
  100Mbps-speed mode, full-duplex mode
  Link speed type is autonegotiation, link duplex type is autonegotiation
  Flow-control is not enabled
  The Maximum Frame Length is 1552
  Broadcast MAX-ratio: 100%
  PVID: 1
  Mdi type: auto
  Port link-type: trunk
   VLAN passing  : 1(default vlan), 10, 20
   VLAN permitted: 1(default vlan), 2-4094
   Trunk port encapsulation: IEEE 802.1q
  Port priority: 0
  Last 300 seconds input:  0 packets/sec 97 bytes/sec
  Last 300 seconds output: 0 packets/sec 74 bytes/sec
  Input (total):  - packets, - bytes
         - broadcasts, - multicasts
  Input (normal): 202712 packets, 31443627 bytes
         179451 broadcasts, 23261 multicasts
  ---- More ----
```

从以上信息可知，交换机 Switch_A 的 VLAN 10 中包含了端口 Ethernet1/0/24，且数据帧是以带有标签（tagged）的形式通过端口。端口 Ethernel/0/24 类型是 Trunk 链路类型，缺省状态（pvid）为 VLAN 1，允许所有 VLAN（1～4094）通过，但实际上只有

VLAN 1、VLAN 10 和 VLAN 20 的数据帧能通过此端口，因为交换机上仅有 VLAN 1、VLAN 10 和 VLAN 20。Switch_B 上 VLAN 的信息和端口 Ethernet 1/0/24 的信息跟 Switch_A 上的信息类似。

（3）连通性测试

在主机 PCA 上 Ping 主机 PCC，结果如下所示：

```
C:\Documents and Settings\Switch_pca>ping 192.168.10.3
 Pinging 192.168.10.3 with 32 bytes of data:
     Reply from 192.168.10.3: bytes=32 time<1ms TTL=128
     Reply from 192.168.10.3: bytes=32 time<1ms TTL=128
     Reply from 192.168.10.3: bytes=32 time<1ms TTL=128
     Reply from 192.168.10.3: bytes=32 time<1ms TTL=128
     ……
```

从以上信息可知，跨交换机同 VLAN 号中的主机通过 Trunk 链路类型端口能够互通。跨交换机不同 VLAN 号中的主机通过 Trunk 链路类型端口不能够互通。由此实现了跨交换机 VLAN 划分。

3．实验任务三：Hybrid 链路类型端口配置

（1）在 Switch_A 上配置端口的链路类型为 Hybrid 类型

配置端口 Ethernet 1/0/1：

```
[Switch_A]interface Ethernet 1/0/1
[Switch_A-Ethernet1/0/1]port link-type hybrid
[Switch_A-Ethernet1/0/1]port hybrid vlan 10 20 untagged
Please wait... Done.
[Switch_A-Ethernet1/0/1]port hybrid vlan 1 tagged
Please wait... Done.
```

配置端口 Ethernet 1/0/10：

```
[Switch_A]interface Ethernet 1/0/10
[Switch_A-Ethernet1/0/10]port link-type hybrid
[Switch_A-Ethernet1/0/10]port hybrid vlan 10 20 untagged
Please wait... Done.
[Switch_A-Ethernet1/0/10]port hybrid vlan 1 tagged
Please wait... Done.
```

查看端口配置信息，如下所示：

```
 [Switch_A]display interface Ethernet 1/0/1
……
PVID: 10
 Mdi type: auto
 Link delay is 0(sec)
 Port link type: hybrid
  Tagged   VLAN ID : 1
  Untagged VLAN ID : 10, 20
```

```
……
//该端口允许VLAN10和VLAN20的数据帧不带标签通过。
[Switch_A]display interface Ethernet 1/0/10
……
PVID: 20
 Mdi type: auto
 Link delay is 0(sec)
 Port link-type: hybrid
  Tagged   VLAN ID : 1
  Untagged VLAN ID : 10, 20
……
```
//该端口允许VLAN10和VLAN20的数据帧不带标签通过。

（2）连通性测试

在主机 PCA 上 Ping 主机 PCB,结果如下所示:

```
C:\Documents and Settings\Switch_pca>ping 192.168.10.2
Pinging 192.168.10.2 with 32 bytes of data:
Reply from 192.168.10.2: bytes=32 time<1ms TTL=128
Reply from 192.168.10.2: bytes=32 time<1ms TTL=128
Reply from 192.168.10.2: bytes=32 time<1ms TTL=128
Reply from 192.168.10.2: bytes=32 time<1ms TTL=128
……
```

由此可知，Hybrid 链路类型端口可以实现不同 VLAN 之间主机的互相访问。

六、实验思考

① 分析当交换机之间连接端口 Ethernet1/0/24 是 Access 链路类型端口时，PCA 与 PCC 分别属于 Switch_A Switch_B 的 VLAN 10，它们是否能通信？

② 分析当交换机之间连接端口 Ethernet1/0/9 分别属于 Switch_A、Switch_B 的 VLAN 10，而且是 Access 链路类型端口时，PCA 与 PCC 也分别属于 Switch_A、Switch_B 的 VLAN 10，它们是否能通信？

实验八　IP 路由基础及 VLAN 间通信

一、实验目的

① 理解路由转发的基本原理和路由的生成；
② 熟悉路由器接口 IP 地址配置；
③ 掌握查看路由表的基本命令；
④ 了解路由表中的路由项要素及含义；
⑤ 掌握 VLAN 间通信路由实现方法。

二、实验内容

① 按图 8-1 所示连接实验设备，规划并配置接口地址；
② 查看、分析路由表项，并进行连通性测试；
③ 利用路由器实现不同 VLAN 间主机通信；
④ 利用三层交换机路由功能实现不同 VLAN 间主机通信。

三、实验设备及组网图

1. 实验设备

一台 H3C 路由器（包括 30-20 或 MSR20 系列路由器），一台交换机，一条专用 Console 配置线缆，两台计算机，超级终端或 SecureCRT 软件。

2. 组网图

通过配置实现主机到网关连通，同时实现主机之间互通，如图 8-1～图 8-3 所示。

四、实验相关知识

1. 路由概念

路由是指通过相互连接的网络把信息从源地点移动到目标地点的活动过程,当路由器收到 IP 信息包时，必须根据 IP 信息包的目的地址，从路由表中选择一条合适的路由记录，即转送此 IP 信息包的最佳路径，然后按路径所指定的网络接口，将 IP 信息包转送出去。路由工作包含两个基本的动作：①确定最佳路径，②网络间交换数据信息。

路由表或路由择域信息库（RIB）是一个存储在路由器或者联网计算机中的电子表格文件或类数据库，其中的每一条路由记录记载了通往每个节点或网络的路径。在有些情况下，还记录路径的路由度量值。路由表中常见表项如图8-4所示。

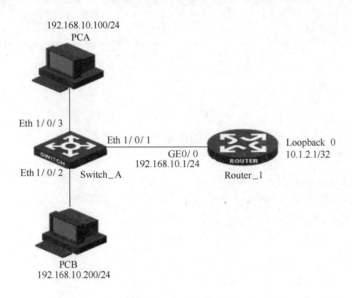

图8-1 IP地址配置与路由表查看实验组网

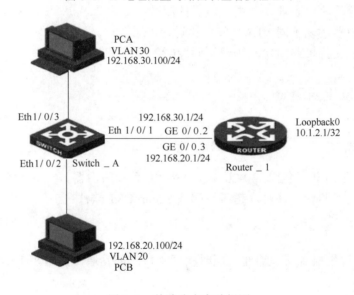

图8-2 单臂路由实验组网

网络地址（network destination）/网络掩码（Netmask）：指本路由器能够到达的网络，它由网络地址与掩码两部分构成。

协议（Protocol）：表示该记录是由某种协议产生的路由，包括直连路由、静态路由、动态路由及引入路由。

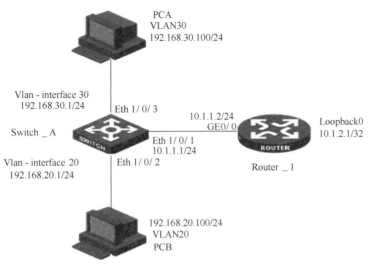

图 8-3 三层交换实验组网

```
[Router] display ip routing-table
Routing Tables: Public
         Destinations: 7        Routes: 7

Destination/Hask    Proto  Pre  Cost   NextHop         Interface
0.0.0.0/0           Static  60   0     10.153.43.1     Eth0/0        ← 缺省路由
1.1.1.0/24          Static  60   0     2.2.2.2         Eth0/0        ← 静态路由
2.2.2.2/32          RIP    100   4     10.153.43.10    Eth0/0        ← 动态路由
3.3.3.3/24          O_ASE  150   1     10.153.43.10    Eth0/0        ← 引入路由
10.153.43.0/24      Direct   0   0     10.153.43.116   Eth0/0
10.153.43.116/32    Direct   0   0     127.0.0.1       InLoop0       ← 直连路由
127.0.0.0/8         Direct   0   0     127.0.0.1       InLoop0
```

图 8-4 路由表

优先级（Preference）：不同路由算法产生的路由优先级别不一样，一般优先级的值越小表示该路由算法越优先，如表 8-1 所示。如果到相同目的地址有多种路由来源，则按不同类型路由的优先级进行选择，优先级最高的路由被添加进路由表。

接口（Interface）：指本路由器输出接口。

下一跳（NextHop）/网关（Gateway）：指下一跳路由器地址或本网网关的 IP 地址。

度量值：有时是跳数（metric），有时是代价值（cost）。

表 8-1 H3C 路由优先级

路由类型	默认优先级
直连路由	0
OSPF 内部路由	10
静态路由	60
RIP 路由	100
OSPF 外部路由	150
BGP 路由	256

2. 路由表中路由的来源

路由的来源包括直连路由（DirectRoute）、静态路由（StaticRoute）和动态路由（DynamicRoute）。另外还可从其他路由域或协议进程中引入路由。

（1）直连路由

直连路由是由链路层协议发现的，一般指路由器接口地址所在网段的路径，该路径信息不需要网络管理员维护，也不需要路由器通过某种算法进行计算获得，只要该接口处于活动状态，路由器就会把通向该网段的路由信息填写到本路由表中，包括 127.0.0.0 网段的环回测试路由。直连路由不能获取与其不直接相连的路由信息。

（2）静态路由

由管理员网络拓扑结构在路由器上进行手工配置的固定的路由。静态路由允许对路由的行为进行精确控制，减少了网络流量。静态路由是在路由器中设置的固定路由，除非网络管理员干预，否则静态路由不会发生变化。由于静态路由不能对网络的改变作出反应，一般用于网络规模不大、拓扑结构固定的网络中。静态路由的优点是简单、高效、可靠，当动态路由与静态路由发生冲突时，以静态路由为准。

1）缺省路由

缺省路由是静态路由的一个特例，也需要人工配置。互联网上有太多的网络和子网，受路由表大小的限制，路由器不可能也没必要为互联网上所有网络和子网指明路径。凡是在路由表中无法查到的目标网络，在路由表中明确指定一个出口，这种路由方法称为缺省路由。

2）黑洞路由

黑洞路由也是静态路由的特殊应用，需要人工配置。是指凡是匹配该路由条目的数据包都将被丢弃，就像宇宙中的黑洞一样，吞噬着所有匹配该路由条目的数据包。

（3）动态路由

由动态路由协议进程自动生成的路由。动态路由是网络中的路由器进程之间相互通信、传递路由信息、利用收到的路由信息更新路由器表的过程。它能实时地适应网络结构的变化。如果路由更新信息表明发生了网络变化，路由选择程序就会重新计算路由，并发出新的路由更新信息。这些信息通过各个网络引起各路由器重新启动其路由算法，并更新各自的路由表，以动态地反映网络拓扑变化。动态路由适用于网络规模大、网络拓扑复杂的网络。当然，各种动态路由协议会不同程度地占用网络带宽和 CPU 资源。

3. 路由表匹配规则

一台路由器上可以同时运行多个路由协议。不同的路由协议都有自己的标准来衡量路由的好坏，并且每个路由协议都把自己认为最好的路由送到路由表中。这样到达一个同样的目的地址，可能有多条分别由不同路由选择协议学习来的不同路由。这就需按一定路由匹配原则来选择具体最佳路由。

路由选择的依据包括目的地址、最长匹配、协议优先级和度量值。一般路由匹配的流程是：从收到的数据包首部提取目的 IP 地址，用每个表项掩码和目的 IP 地址逐位相"与"，获取目的网络，看结果是否和相应表项的网络地址匹配，如没有匹配表项，则按缺省路由转发或丢弃；如有一条或多条路由表项匹配，则按掩码最精确匹配的路由优先；如果有多条路由符合最长匹配原则，则比较协议优先级，优先级小的路由优先选择；如果优先级相同，在比较度量值，度量值小的路由优先选择。

路由选择原则有以下 5 点。

① 首先根据目的地址和最长匹配原则进行查找。所谓的最长匹配就是路由查找时，使用路由表中到达同一目的地的子网掩码最长的路由。

图 8-5　最长匹配路由选择原则

② 若有两条或两条以上路由符合，则按协议优先值越小，优先级越高的原则选择。

图 8-6　协议优先级路由选择原则

③ 当协议优先级相同时，按度量值越小，优先级越高的原则选择。

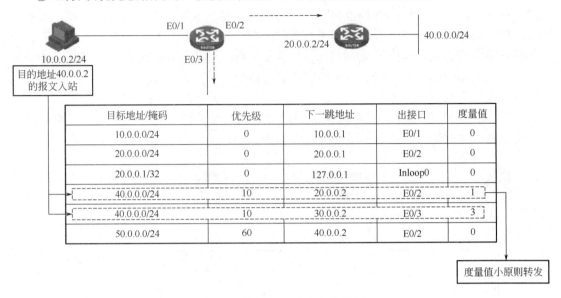

图 8-7 度量值小路由选择原则

④ 如路由各表项不能匹配数据包目的网络，并且路由表中有缺省路由时，则把数据报传送给缺省路由器。

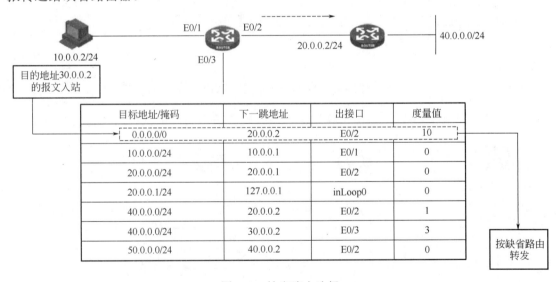

图 8-8 缺省路由选择

⑤ 如路由器没有默认路由，则丢弃该数据包。

4．不同 VLAN 间通信路由的实现

引入 VLAN 之后，每个交换机可被划分成多个 VLAN,而每个 VLAN 对应一个 IP 网段。VLAN 隔离广播域，不同 VLAN 之间是二层隔离的，即不同 VLAN 的主机发出的数据帧被交换机内部隔离了。但建网的最终目的是实现网络的互联互通，所以还需要相应方案来实现不同 VLAN 间的通信。VLAN 间实现通信方案主要有三种。

（1）多个端口路由器实现 VLAN 间通信

最传统的方法是将二层交换机与路由器结合，通过路由器多个端口连接不同 VLAN 的接口，交换机与路由器连接的端口都工作在 Access 模式下。

如图 8-9 所示，路由器分别使用 3 个工作在路由模式的以太网接口连接到交换机的 3 个不同 VLAN 中，不同的 VLAN 对应不同的子网段，路由器连接不同 VLAN 相应接口地址是各 VLAN 内各主机的网关。当 VLAN 10 中源主机 A 需向 VLAN 20 中目的主机 B 发送信息时，主机 A 与主机 B 不在同一子网内，数据按远程数据转发给 VLAN 10 所在子网的网关。首先交换机 VLAN 10 内 1 端口将主机 A 发送的数据帧打上 VLAN 10 标签，从交换电路内部转发给同 VLAN 的交换机 2 端口，工作在 Access 模式的 2 端口剥离 VLAN 10 标签，将数据包发送到路由器 1 端口。路由器收到不带 VLAN 标签的数据包，根据数据包的目的 IP 地址查路由表确定目的转发接口，然后将数据帧通过路由器 2 端口转发给交换机 VLAN 20 中 7 端口；7 端口打上 VLAN 20 帧标签，从交换电路转发给交换机 8 端口，最后，交换机 8 端口剥离数据帧二层 VLAN 20 标签，并转发至目的主机 B。由此实现不同 VLAN 间主机相互通信。

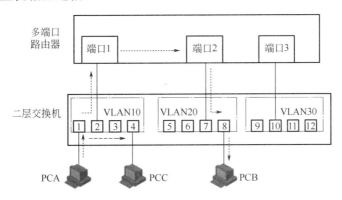

图 8-9　多个端口路由器实现 VLAN 间通信

（2）单臂路由实现 VLAN 间通信

单臂路由是为了避免物理端口的浪费，通过一条物理线连接路由器，利用交换机端口工作在 Trunk 模式下时可以允许多个 VLAN 帧通过的数据转发机制来实现多个 VLAN 互通的路由简化技术。它要求路由器支持 802.1Q 封装和子接口技术，能识别二层 VLAN 标签并能剥离二层 VLAN 标签，同时，路由器的一个接口上通过配置子接口（或"逻辑接口"，并不存在真正的物理接口）方式，实现原来相互隔离的不同 VLAN 之间的互联互通。

如图 8-10 所示，交换机通过 Trunk 模式的 13 端口连接路由器以太网端口，配置交换机 13 端口允许 VLAN 10、VLAN 20、VLAN 30 数据帧通过。在路由器的端口上创建三个子接口，每个子接口配置属于相应 VLAN 网段的 IP 地址及可识别的 VLAN 标签值，允许接收 VLAN 10、VLAN 20、VLAN 30 数据帧。

如果 VLAN 10 内源主机 A 向 VLAN 20 内目的主机 B 发送信息，打帧标记为 VLAN 10 的数据帧会由交换机 Trunk 端口 13 转发到路由器的 Eth0/0.1 子接口，在路由器子接口上剥离 VLAN 10 标签，再按路由表选择路径,确定目的接口为 Eth0/0.2 子接口，Eth0/0.2 子接口重新打帧标记为 VLAN 20，帧标记为 VLAN 20 的数据帧由路由器的 Eth0/0.2 接口转发给交换机 13 端口，再由交换电路转发至 VLAN 20 内的 8 端口，交换机 8 端口剥离数据帧二层 VLAN 20 标签后，转发到目的主机 B。

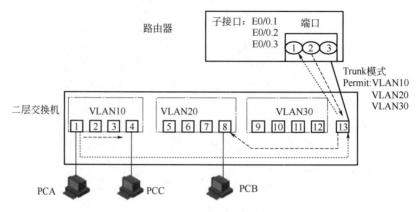

图 8-10　单臂路由实现 VLAN 间通信

采用"单臂路由"方式进行 VLAN 间路由时数据帧需要在 Trunk 链路上往返发送，从而引起一定的转发延迟，如果 VLAN 间路由数据量较大，会消耗路由器大量 CPU 和内存资源，造成转发性能的瓶颈。

（3）三层交换机实现 VLAN 间通信

三层交换是指在二层交换机中嵌入路由模块而取代传统路由器实现交换与路由相结合的网络技术，它利用三层交换机的路由模块识别数据包 IP 地址功能，查找路由表进行选路转发。现有园区网内部主要采用三层交换技术实现不同 VLAN 间的通信。

在三层交换机上跨 VLAN 间路由时，需要创建交换虚拟接口（switch virtual interface，SVI）。交换虚拟接口是在三层转发引擎和二层转发引擎上建立的逻辑接口，功能与路由器的接口相似，可以配置 IP 地址。一个 VLAN 只能创建一个虚拟接口，虚拟接口的 IP 地址是该 VLAN 中各主机的网关。

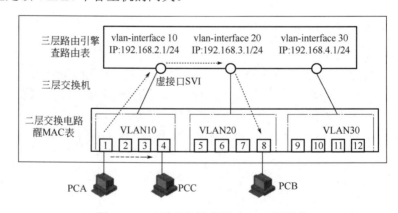

图 8-11　三层交换机实现 VLAN 间通信

如图 8-11 所示，如支持 IP 协议的 VLAN 10 内主机 A 与 VLAN 20 内主机 B 须通过第三层交换机进行通信，首先主机 A 在发送第一个数据包时，把源主机 A 的 IP 地址与目的主机 B 的 IP 地址比较，判断 B 站是否与主机 A 在同一子网内，若主机 B 与主机 A 在同一子网内，则进行二层的转发；若两个站点不在同一子网内，将按远程通信进行转发，须将该数据包发给本网网关，VLAN 10 的网关是三层交换机路由模块上的虚拟接口，因此主机 A 向三层交换机的三层交换模块发出 ARP 封包。三层交换模块解析发送主机 A 的目的 IP 地址，向目的 IP 地址所在网段的 VLAN 20 发送 ARP 请求。主机 B 得到此 ARP 请求后，向三层交换模块回复其 MAC 地址，三层交换模块保存此地址并回复发送主机 A，同时将主机 B 的 MAC 地址发送到二层交换引擎的 MAC 地址表中，此后，主机 A 向主机 B 发送的后续数据包便全部按二层交换处理（称为一次路由多次交换），信息得以高速交换。由于仅仅在路由过程中需要三层处理，绝大部分数据都通过二层交换转发，三层交换机的速度很快，接近二层交换机的速度。

五、实验内容

1. 实验任务一：路由器端口 IP 地址配置和查看接口、路由表信息

（1）搭建实验环境，初始化实验设备

按图 8-1 所示搭建实验环境，清除配置文件，以出厂配置重启网络设备。

```
<H3C>reset saved-configuration
<H3C>reboot
......
<H3C>
```

（2）按表 8-2 所示更改设备名称并配置主机、路由器端口 IP 地址

表 8-2　IP 地址规划表

设备名称	接口	IP 地址	网关
Router_1	GE0/0	192.168.10.1/24	
	Loopback0	10.1.2.1/32	
PCA		192.168.10.100	192.168.10.1
PCB		192.168.10.200	192.168.10.1

```
[Router_1]display ip routing-table

Routing Tables: Public
       Destinations : 2         Routes : 2
Destination/Mask    Proto  Pre  Cost        NextHop         Interface
127.0.0.0/8         Direct 0    0           127.0.0.1       InLoop0
127.0.0.1/32        Direct 0    0           127.0.0.1       InLoop0
```

由以上输出可知，路由器初始化后路由表中只有环回地址的直连路由。

配置 Router_1：

```
[Router_1]interface GE0/0
[Router_1-GE0/0]ip address 192.168.10.1 24           //路由器以太网接口配置 IP 地
```

址。

[Router_1]**display ip routing-table**   //配置完成后再次查看路由表。
Routing Tables: Public
         Destinations : 4   Routes : 4

Destination/Mask   Proto   Pre   Cost   NextHop   Interface

127.0.0.0/8   Direct   0   0   127.0.0.1   InLoop0
127.0.0.1/32   Direct   0   0   127.0.0.1   InLoop0
192.168.10.0/24   Direct   0   0   192.168.10.1   Eth0/0
192.168.10.1/32   Direct   0   0   127.0.0.1   InLoop0

[Router_1]**display brief interface**   //查看路由器接口摘要信息。
The brief information of interface(s) under route mode:
Interface   Link   Protocol-link   Protocol type   Main IP
Aux0   DOWN   DOWN   --   --
GE0/0   UP   UP   ETHERNET   192.168.10.1
GE0/1   DOWN   DOWN   ETHERNET   --
NULL0   UP   UP(spoofing)   NULL   --
S6/0   DOWN   DOWN   PPP   --
S6/1   DOWN   DOWN   PPP   --

[Router_1]**display interface GE0/0**   //查看路由器以太网接口详细信息。
GigabitEthernet0/0 current state: UP
Line protocol current state: UP
Description: GigabitEthernet0/0 Interface
The Maximum Transmit Unit is 1500, Hold timer is 10(sec)
Internet Address is 192.168.10.1/24 Primary
IP Packet Frame Type: PKTFMT_ETHNT_2, Hardware Address: 000f-e2a2-c709
Ipv6 Packet Frame Type: PKTFMT_ETHNT_2, Hardware Address: 000f-e2a2-c709
Media type is twisted pair, loopback not set, promiscuous mode not set
100Mb/s, Full-duplex, link type is autonegotiation
Output flow-control is disabled, input flow-control is disabled
Output queue : (Urgent queuing : Size/Length/Discards) 0/100/0
Output queue : (Protocol queuing : Size/Length/Discards) 0/500/0
Output queue : (FIFO queuing : Size/Length/Discards) 0/75/0
Last clearing of counters: Never
    Last 300 seconds input rate 116.86 bytes/sec, 934 位 s/sec, 1.19 packets/sec
    Last 300 seconds output rate 0.60 bytes/sec, 4 位 s/sec, 0.01 packets/sec
      Input: 508 packets, 49169 bytes, 508 buffers
            431 broadcasts, 77 multicasts, 0 pauses
            0 errors, 0 runts, 0 giants
            0 crc, 0 align errors, 0 overruns
            0 dribbles, 0 drops, 0 no buffers
      Output:3 packets, 180 bytes, 3 buffers

```
            3 broadcasts, 0 multicasts, 0 pauses

[Router_1]display interface S6/0            //查看路由器串口详细信息。
Serial6/0 current state: DOWN
Line protocol current state: DOWN
Description: Serial6/0 Interface
The Maximum Transmit Unit is 1500, Hold timer is 10(sec)
Internet protocol processing : disabled
Link layer protocol is PPP
LCP initial
Output queue : (Urgent queuing : Size/Length/Discards)  0/100/0
Output queue : (Protocol queuing : Size/Length/Discards) 0/500/0
Output queue : (FIFO queuing : Size/Length/Discards)  0/75/0
Physical layer is synchronous, Baudrate is 64000 bps
Interface is DCE, Cable type is V24, Clock mode is DCECLK
Last clearing of counters: Never
    Last 300 seconds input rate 0.00 bytes/sec, 0 位s/sec, 0.00 packets/sec
    Last 300 seconds output rate 0.00 bytes/sec, 0 位s/sec, 0.00 packets/sec
    Input:  0 packets, 0 bytes
            0 broadcasts, 0 multicasts
            0 errors, 0 runts, 0 giants
            0 CRC, 0 align errors, 0 overruns
            0 dribbles, 0 aborts, 0 no buffers
            0 frame errors
    Output: 0 packets, 0 bytes
            0 errors, 0 underruns, 0 collisions
 ---- More ----


[Router_1]interface GE0/0
[Router_1- GE0/0]shutdown         //在 Router_1 上关闭接口 GE0/0。
 [Router_1]display ip routing-table       //查看路由表。
Routing Tables: Public
       Destinations : 2      Routes : 2
Destination/Mask   Proto  Pre  Cost      NextHop         Interface
127.0.0.0/8        Direct  0    0        127.0.0.1       InLoop0
127.0.0.1/32       Direct  0    0        127.0.0.1       InLoop0
```

在接口 shutdown 之后，所运行的链路层协议关闭，直连路由也就自然消失了。

```
[Router_1]interface GE 0/0
[Router_1- GE0/0]undo shutdown      // 再激活接口。
```

等到链路层协议 up 以后，再次查看路由表，发现 Eth0/0 接口的直连路由又出现了。

```
[Router_1- GE0/0]
 %Apr  3 11:40:51:299 2010 Router_1 DRVMSG/1/DRVMSG:       Ethernet0/0: change
status to up
 %Apr  3 11:40:51:299 2010 Router_1 IFNET/4/UPDOWN:
  Line protocol on the interface Ethernet0/0 is UP
```

主机 PCA 到网关的连通性测试。

```
C:\Documents and Settings\Administrator>ping 192.168.1.1
 Pinging 192.168.1.1 with 32 bytes of data:
    Reply from 192.168.1.1: bytes=32 time<1ms TTL=255
    Reply from 192.168.1.1: bytes=32 time<1ms TTL=255
    Reply from 192.168.1.1: bytes=32 time<1ms TTL=255
    Reply from 192.168.1.1: bytes=32 time<1ms TTL=255

Ping statistics for 192.168.1.1:
    Packets: Sent = 4, Received = 4, Lost = 0 (0% loss),
Approximate round trip times in milli-seconds:
Minimum = 0ms, Maximum = 0ms, Average = 0ms
```

主机 PCA 到路由器 Loopback 口的连通性测试。

```
C:\Documents and Settings\Administrator>ping 10.1.2.1
 Pinging 10.1.2.1 with 32 bytes of data:
 Reply from 10.1.2.1: bytes=32 time<1ms TTL=255
 Reply from 10.1.2.1: bytes=32 time<1ms TTL=255
 Reply from 10.1.2.1: bytes=32 time<1ms TTL=255
 Reply from 10.1.2.1: bytes=32 time<1ms TTL=255

Ping statistics for 10.1.2.1:
    Packets: Sent = 4, Received = 4, Lost = 0 (0% loss),
Approximate round trip times in milli-seconds:
    Minimum = 0ms, Maximum = 0ms, Average = 0ms
```

2. 实验任务二：VLAN 间路由

（1）方案一：用路由器单端口实现 VLAN 间的路由——单臂路由

1）搭建实验环境，初始化实验设备

按图 8-2 所示搭建实验环境，清除配置文件，以出厂配置重启网络设备。

```
<H3C>reset saved-configuration
<H3C>reboot
……
<H3C>
```

2）按表 8-3 所示更改设备名称并配置主机、路由器端口 IP 地址

表 8-3 IP 地址规划表

设备名称	接口/vlan	IP 地址	网关
Router_1	GE0/0.2	192.168.20.1/24	
	GE0/0.3	192.168.30.1/24	
	LoopBack0	10.1.2.1/32	

续表

设备名称	接口/vlan	IP 地址	网关
Switch_A	Eth1/0/1		
	Eth1/0/2 （Vlan 2）		
	Eth1/0/3 （Vlan 3）		
PCA		192.168.30.100	192.168.30.1
PCB		192.168.20.100	192.168.20.1

Router_1 配置如下：

```
[Router_1]interface GE0/0.2
[Router_1-GE 0/0.2]vlan-type dot1q vid 2
[Router_1- GE 0/0.2]ip address 192.168.20.1 24
[Router_1- GE 0/0]interface GE0/0.3
[Router_1- GE 0/0.3]vlan-type dot1q vid 3
[Router_1- GE 0/0.3]ip address 192.168.30.1 24
[Router_1- GE 0/0.3]
%Apr  3 12:13:37:879 2010 Router_1 IFNET/4/UPDOWN:
 Line protocol on the interface Ethernet0/0.2 is UP
```

Switch_A 配置如下：

```
[Switch_A]vlan 2
[Switch_A-vlan2]port Ethernet 1/0/2
[Switch_A-Ethernet1/0/2]port link-type access
[Switch_A-Ethernet1/0/2]port access vlan 2
[Switch_A-vlan2]quit
[Switch_A]vlan 3
[Switch_A-vlan3]port Ethernet 1/0/3
[Switch_A-Ethernet1/0/3]port link-type access
[Switch_A-Ethernet1/0/3]port access vlan 3
[Switch_A-vlan3]quit
[Switch_A]interface Ethernet 1/0/1
[Switch_A-Ethernet1/0/1]port link-type trunk
[Switch_A-Ethernet1/0/1]port trunk permit vlan all
 Please wait........................................ Done.
```

此时，PCA 和 PCB 可以通过 Router_1 进行通信。
PCB Ping PCA 网关。

```
C:\Documents and Settings\Administrator>ping 192.168.30.1
 Pinging 192.168.30.1 with 32 bytes of data:
   Reply from 192.168.30.1: bytes=32 time=1ms TTL=63
   Reply from 192.168.30.1: bytes=32 time=1ms TTL=63
   Reply from 192.168.30.1: bytes=32 time=1ms TTL=63
   Reply from 192.168.30.1: bytes=32 time<1ms TTL=63
Ping statistics for 192.168.2.200:
   Packets: Sent = 4, Received = 4, Lost = 0 （0% loss），
```

```
Approximate round trip times in milli-seconds:
    Minimum = 0ms, Maximum = 1ms, Average = 0ms
```

（2）方案二：用三层交换机实现不同 VLAN 间主机的通信

1）搭建实验环境，初始化实验设备

按图 8-3 所示搭建实验环境，清除配置文件，以出厂配置重启网络设备。

```
<H3C>reset saved-configuration
<H3C>reboot
......
<H3C>
```

2）按表 8-4 所示更改设备名称并配置 PC、路由器接口、交换机虚接口 IP 地址

表 8-4　IP 地址规划表

设备名称	接口	IP 地址	网关
Router_1	GE0/0	10.1.1.2/24	
	LoopBack0	10.1.2.1/32	
Switch_A	Eth1/0/1 vlan-interface 1	10.1.1.1/24	
	Vlan-interface 30	192.168.30.1/24	
	Vlan-interface 20	192.168.20.1/24	
PCA		192.168.30.200	192.168.30.1
PCB		192.168.20.200	192.168.20.1

Switch_A 配置如下：

```
[Switch_A]vlan 2
[Switch_A-vlan2]port e1/0/2
[Switch_A-vlan2]quit
[Switch_A]interface Vlan-interface 2
[Switch_A-Vlan-interface2]
%Apr 26 12:08:03:623 2000 Switch_A IFNET/4/LINK UPDOWN:
 Vlan-interface2: link status is UP
[Switch_A-Vlan-interface2]ip address 192.168.2.1 24
[Switch_A-Vlan-interface2]
%Apr 26 12:08:19:35 2000 Switch_A IFNET/4/UPDOWN:
 Line protocol on the interface Vlan-interface2 is UP
[Switch_A-Vlan-interface2]interface vlan-interface 1
[Switch_A-Vlan-interface1]
%Apr 26 12:08:36:589 2000 Switch_A IFNET/4/LINK UPDOWN:
 Vlan-interface1: link status is UP
[Switch_A-Vlan-interface1]ip address 192.168.1.1 24
[Switch_A-Vlan-interface1]
%Apr 26 12:08:53:303 2000 Switch_A IFNET/4/UPDOWN:
 Line protocol on the interface Vlan-interface1 is UP
```

```
[Switch_A]interface Ethernet 1/0/1
[Switch_A-Ethernet1/0/1]port link-mode route
%Apr 26 12:09:29:373 2000 Switch_A IFNET/4/LINK UPDOWN:
 Ethernet1/0/1: link status is DOWN
[Switch_A-Ethernet1/0/1]
%Apr 26 12:09:31:317 2000 Switch_A IFNET/4/LINK UPDOWN:
 Ethernet1/0/1: link status is UP
[Switch_A-Ethernet1/0/1]ip address 10.1.1.1 24
```

Router_1 配置如下：

```
[Router_1]interface Ethernet 0/0
[Router_1-Ethernet0/0]ip address 10.1.1.2 24
```

可以在 Switch_A 上查看路由表，命令如下：

```
[Switch_A]display ip routing-table
Routing Tables: Public
       Destinations : 8        Routes : 8

Destination/Mask     Proto  Pre  Cost        NextHop          Interface

10.1.1.0/24          Direct 0    0           10.1.1.1         Eth1/0/1
10.1.1.1/32          Direct 0    0           127.0.0.1        InLoop0
127.0.0.0/8          Direct 0    0           127.0.0.1        InLoop0
127.0.0.1/32         Direct 0    0           127.0.0.1        InLoop0
192.168.1.0/24       Direct 0    0           192.168.1.1      Vlan1
192.168.1.1/32       Direct 0    0           127.0.0.1        InLoop0
192.168.2.0/24       Direct 0    0           192.168.2.1      Vlan2
192.168.2.1/32       Direct 0    0           127.0.0.1        InLoop0
```

测试 PC 到网关的可达性。在 PCA 上用 Ping 命令测试 PCB 的可达性。

```
C:\Documents and Settings\Administrator>ping 192.168.1.1
  Pinging 192.168.1.1 with 32 bytes of data:
    Reply from 192.168.1.1: bytes=32 time=2ms TTL=255
    Reply from 192.168.1.1: bytes=32 time=1ms TTL=255
    Reply from 192.168.1.1: bytes=32 time=1ms TTL=255
    Reply from 192.168.1.1: bytes=32 time=1ms TTL=255

Ping statistics for 192.168.1.1:
    Packets: Sent = 4, Received = 4, Lost = 0 (0% loss),
Approximate round trip times in milli-seconds:
    Minimum = 1ms, Maximum = 2ms, Average = 1ms
```

测试 PC 之间的可达性，例如，在 PCA 上用 Ping 命令测试到 PCB 的可达性。

```
C:\Documents and Settings\Administrator>ping 192.168.2.200
  Pinging 192.168.2.200 with 32 bytes of data:
    Reply from 192.168.2.200: bytes=32 time=26ms TTL=63
    Reply from 192.168.2.200: bytes=32 time<1ms TTL=63
    Reply from 192.168.2.200: bytes=32 time<1ms TTL=63
```

```
    Reply from 192.168.2.200: bytes=32 time<1ms TTL=63

Ping statistics for 192.168.2.200:
    Packets: Sent = 4, Received = 4, Lost = 0 (0% loss),
Approximate round trip times in milli-seconds:
    Minimum = 0ms, Maximum = 26ms, Average = 6ms
```

以上输出信息显示：PCA 和 PCB 之间可以互通。

测试 PC 到 Switch_A 接口 E1/0/1 的连通性，以 PCA 为例：

```
C:\Documents and Settings\Administrator>ping 10.1.1.1
  Pinging 10.1.1.1 with 32 bytes of data:

    Reply from 10.1.1.1: bytes=32 time=1ms TTL=255
    Reply from 10.1.1.1: bytes=32 time=1ms TTL=255
    Reply from 10.1.1.1: bytes=32 time=1ms TTL=255
    Reply from 10.1.1.1: bytes=32 time=1ms TTL=255

Ping statistics for 10.1.1.1:
    Packets: Sent = 4, Received = 4, Lost = 0 (0% loss),
  Approximate round trip times in milli-seconds:
Minimum = 1ms, Maximum = 1ms, Average = 1ms
```

六、实验思考

① 给出路由器某以太网接口和某同/异步串接口的详细参数信息，它们有什么区别？
② 分析路由器路由表信息。
③ 试述单臂路由的优缺点。

实验九　静态路由配置

一、实验目的

① 掌握静态路由配置方法；
② 掌握缺省路由应用方法。

二、实验内容

① 按图 9-1 所示连接实验设备，规划并配置接口地址；
② 配置静态路由，查看分析路由，实现设备互通；
③ 配置缺省路由，查看分析路由，实现设备互通。

三、实验设备及组网图

1．实验设备

三台 H3C 路由器，一条专用 Console 配置线缆，一台计算机，超级终端或 Secure CRT 软件。

2．组网图

如图 9-1 所示，将三台路由器分别用双绞线和串口线相连，配置静态路由，实现不同网段互通。

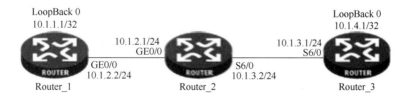

图 9-1　实验组网

四、实验相关知识

静态路由是指由用户或网络管理员手工配置的路由信息。静态路由一般适用于比较简单的网络环境,在这样的环境中,网络管理员易于清楚地了解网络的拓扑结构,便于设置正确的路由信息。

大型和复杂的网络环境通常不宜采用静态路由。一方面,网络管理员难以全面了解整个网络的拓扑结构;另一方面,当网络的拓扑结构和链路状态发生变化时,路由器中的静态路由信息需要大范围地调整,这一工作的难度和复杂程度非常高。当网络发生变化或出现故障时,不能重选路由,很可能使路由失败。

配置路由器的基本思路如下。

第一步:在配置路由器之前,需要将组网需求具体化、详细化,包括组网目的、路由器在网络互连中的角色、子网的划分、广域网类型和传输介质的选择、网络的安全策略和网络可靠性需求等。

第二步:根据以上信息素绘一个清晰完整的组网图,规划路由器各接口的 IP 地址。

第三步:配置路由器的广域网接口。首先根据选择的广域网传输介质配置接口的物理工作参数(如串口的同/异步、波特率和同步时钟等);对于拨号口,还需要配置 DCC 参数;然后根据选择的广域网类型,配置接口封装的链路层协议及相应的工作参数。

第四步:根据子网的划分,配置路由器各接口的 IP 地址。

第五步:配置路由,如果需要启动动态路由协议,还需配置相关动态路由协议的工作参数。

第六步:如果有特殊的安全需求,则需进行路由器的安全性配置。

第七步:如果有特殊的可靠性需求,则需进行路由器的可靠性配置。

五、实验内容

1. 实验任务一:静态路由配置

(1)搭建实验环境,初始化实验设备

按图 9-1 所示搭建实验环境,清除配置文件,以出厂配置重启网络设备。

```
<H3C>reset saved-configuration
<H3C>reboot
……
<H3C>
```

(2)按表 9-1 所示更改设备名称并配置路由器端口 IP 地址

表 9-1 IP 地址规划表

设备名称	接口	IP 地址
Router_1	LoopBack0	10.1.1.1/32
	GE0/0	10.1.2.1/24
Router_2	GE0/0	10.1.2.2/24
	S6/0	10.1.3.1/24
Router_3	S6/0	10.1.3.2/24
	LoopBack0	10.1.4.1/32

配置 Router_1:

```
[Router_1]interface GE 0/0
[Router_1-GE 0/0]ip address 10.1.2.1 24
[Router_1]interface LoopBack 0
[Router_1-LoopBack0]ip address 10.1.1.1 32
```

配置 Router_2:

```
[Router_2]interface GE 0/0
[Router_2-GE 0/0]ip address 10.1.2.2 24
[Router_2]interface Serial 6/0
[Router_2-Serial1/0]ip address 10.1.3.1 24
```

配置 Router_3:

```
[Router_3]interface Serial 6/0
[Router_3-Serial1/0]ip address 10.1.3.2 24
[Router_3]interface LoopBack 0
[Router_3-LoopBack0]ip address 10.1.4.1 32
```

（3）配置静态路由

在 Router_1 和 Router_3 上配置静态路由。

配置 Router_1:

```
[Router_1]ip route-static 10.1.3.1 24 10.1.2.2
[Router_1]ip route-static 10.1.4.1 24 10.1.2.2
```

配置 Router_3:

```
[Router_3]ip route-static 10.1.2.0 24 Serial 1/0 10.1.3.1
[Router_3]ip route-static 10.1.1.0 24 10.1.3.1
```

如果出接口是串口，配置时既可以写下一跳地址，也可以写出接口，配置完成后，在路由器上查看路由表。例如，在 Router_1 上查看路由表：

```
[Router_1]display ip routing-table
Routing Tables: Public
        Destinations : 7       Routes : 7
Destination/Mask      Proto  Pre  Cost      NextHop        Interface

10.1.1.1/32           Direct 0    0         127.0.0.1      InLoop0
10.1.2.0/24           Direct 0    0         10.1.2.1       Eth0/0
10.1.2.1/32           Direct 0    0         127.0.0.1      InLoop0
10.1.3.0/24           Static 60   0         10.1.2.2       GE0/0
10.1.4.0/24           Static 60   0         10.1.2.2       GE0/0
127.0.0.0/8           Direct 0    0         127.0.0.1      InLoop0
127.0.0.1/32          Direct 0    0         127.0.0.1      InLoop0
```

测试 Router_1 和 Router_3 之间的可达性。在 Router_1 上用 Ping 命令测试到 Router_3 的 S1/0 口的可达性。

```
[Router_1]ping 10.1.3.2
```

```
  PING 10.1.3.2: 56  data bytes, press CTRL_C to break
    Reply from 10.1.3.2: bytes=56 Sequence=1 ttl=254 time=27 ms
    Reply from 10.1.3.2: bytes=56 Sequence=2 ttl=254 time=27 ms
    Reply from 10.1.3.2: bytes=56 Sequence=3 ttl=254 time=28 ms
    Reply from 10.1.3.2: bytes=56 Sequence=4 ttl=254 time=27 ms
    Reply from 10.1.3.2: bytes=56 Sequence=5 ttl=254 time=28 ms
  --- 10.1.3.2 ping statistics ---
    5 packet(s) transmitted
    5 packet(s) received
    0.00% packet loss
    round-trip min/avg/max = 27/27/28 ms
```

表明可以 Ping 通。

在 Router_1 上 Ping 10.1.4.1：

```
[Router_1]ping 10.1.4.1
  PING 10.1.4.1: 56  data bytes, press CTRL_C to break
    Request time out
    Request time out
    Request time out
    Request time out
    Request time out

  --- 10.1.4.1 ping statistics ---
    5 packet(s) transmitted
    0 packet(s) received
    100.00% packet loss
```

或者是在 Router_1 上使用 ping –a 10.1.1.1 10.1.3.2；

```
[Router_1]ping -a 10.1.1.1 10.1.3.2
  PING 10.1.3.2: 56  data bytes, press CTRL_C to break
    Request time out
    Request time out
    Request time out
    Request time out
    Request time out

  --- 10.1.3.2 ping statistics ---
    5 packet(s) transmitted
    0 packet(s) received
    100.00% packet loss
```

发现此时不能 Ping 通。

```
  [Router2]display ip routing-table
Routing Tables: Public
        Destinations : 7       Routes : 7

Destination/Mask     Proto  Pre  Cost       NextHop        Interface

10.1.2.0/24          Direct 0    0          10.1.2.2       GE0/0
```

```
10.1.2.2/32         Direct  0   0   127.0.0.1   InLoop0
10.1.3.0/24         Direct  0   0   10.1.3.1    S6/0
10.1.3.1/32         Direct  0   0   127.0.0.1   InLoop0
10.1.3.2/32         Direct  0   0   10.1.3.2    S6/0
127.0.0.0/8         Direct  0   0   127.0.0.1   InLoop0
127.0.0.1/32        Direct  0   0   127.0.0.1   InLoop0
```

通过查看路由表，发现 Router_2 的路由表中没有到 10.1.1.1/32 和 10.1.1.4.1/32 的路由。因此，在 Router_2 上添加静态路由。

```
[Router_2]ip route-static 10.1.1.1 32 10.1.2.1
[Router_2]ip route-static 10.1.4.1 32 10.1.3.2
[Router_1]ping -a 10.1.1.1 10.1.4.1
  PING 10.1.4.1: 56  data bytes, press CTRL_C to break
    Reply from 10.1.4.1: bytes=56 Sequence=1 ttl=254 time=25 ms
    Reply from 10.1.4.1: bytes=56 Sequence=2 ttl=254 time=26 ms
    Reply from 10.1.4.1: bytes=56 Sequence=3 ttl=254 time=26 ms
    Reply from 10.1.4.1: bytes=56 Sequence=4 ttl=254 time=26 ms
    Reply from 10.1.4.1: bytes=56 Sequence=5 ttl=254 time=26 ms

  --- 10.1.4.1 ping statistics ---
    5 packet(s) transmitted
    5 packet(s) received
    0.00% packet loss
    round-trip min/avg/max = 25/25/26 ms
```

可见实现了各网段之间的互通。

2. 实验任务二：配置缺省路由

在路由器上合理地配置缺省路由能够减少路由表中的表项数量，节省路由表空间，加快路由匹配速度。

默认路由经常在末端网络中使用。末端网络是指仅有一个出口连接外部的网络。在 Router_1 和 Router_3 上用 LoopBack 口模拟末端网络，就可以在 Router_1 和 Router_3 上配置默认路由。

（1）建立物理连接

实验组网如图 9-1 所示，IP 地址配置如表 9-1 所示。

（2）配置默认路由

Router_1 配置：

`[Router_1]ip route-static 0.0.0.0 0 10.1.2.2`

Router_2 配置：

```
[Router_2]ip route-static 10.1.1.1 32 10.1.2.2
                                    //配置的下一跳地址是路由器自身。
                                    // 接口地址，报错!
Error: Invalid Nexthop Address.
[Router_2]ip route-static 10.1.1.1 32 10.1.2.1
```

[Router_2]**ip route-static 10.1.4.1 32 10.1.3.2**

Router_3 配置：

[Router_3]**ip route-static 0.0.0.0 0 Serial 6/0 10.1.3.1**

测试各路由器之间的连通性，在 Router_1 上使用 Ping 命令测试：

[Router_1]**ping 10.1.3.2**
  PING 10.1.3.2: 56 data bytes, press CTRL_C to break
    Reply from 10.1.3.2: bytes=56 Sequence=1 ttl=254 time=26 ms
    Reply from 10.1.3.2: bytes=56 Sequence=2 ttl=254 time=26 ms
    Reply from 10.1.3.2: bytes=56 Sequence=3 ttl=254 time=26 ms
    Reply from 10.1.3.2: bytes=56 Sequence=4 ttl=254 time=26 ms
    Reply from 10.1.3.2: bytes=56 Sequence=5 ttl=254 time=26 ms

  --- 10.1.3.2 ping statistics ---
    5 packet(s) transmitted
    5 packet(s) received
    0.00% packet loss
    round-trip min/avg/max = 26/26/26 ms

[Router_1]**ping 10.1.4.1**
  PING 10.1.4.1: 56 data bytes, press CTRL_C to break
    Reply from 10.1.4.1: bytes=56 Sequence=1 ttl=254 time=26 ms
    Reply from 10.1.4.1: bytes=56 Sequence=2 ttl=254 time=25 ms
    Reply from 10.1.4.1: bytes=56 Sequence=3 ttl=254 time=25 ms
    Reply from 10.1.4.1: bytes=56 Sequence=4 ttl=254 time=25 ms
    Reply from 10.1.4.1: bytes=56 Sequence=5 ttl=254 time=26 ms

  --- 10.1.4.1 ping statistics ---
    5 packet(s) transmitted
    5 packet(s) received
    0.00% packet loss
    round-trip min/avg/max = 25/25/26 ms

[Router_1]**ping -a 10.1.1.1 10.1.4.1**
  PING 10.1.4.1: 56 data bytes, press CTRL_C to break
    Reply from 10.1.4.1: bytes=56 Sequence=1 ttl=254 time=25 ms
    Reply from 10.1.4.1: bytes=56 Sequence=2 ttl=254 time=25 ms
    Reply from 10.1.4.1: bytes=56 Sequence=3 ttl=254 time=26 ms
    Reply from 10.1.4.1: bytes=56 Sequence=4 ttl=254 time=26 ms
    Reply from 10.1.4.1: bytes=56 Sequence=5 ttl=254 time=26 ms

  --- 10.1.4.1 ping statistics ---
    5 packet(s) transmitted
    5 packet(s) received
    0.00% packet loss
    round-trip min/avg/max = 25/25/26 ms

[Router_1]**display ip routing-table**   // 查看 Router_1 路由表。
Routing Tables: Public

```
         Destinations : 8        Routes : 8

Destination/Mask      Proto   Pre  Cost       NextHop         Interface
0.0.0.0/0             Static  60   0          10.1.2.2        GE0/0
10.1.1.1/32           Direct  0    0          127.0.0.1       InLoop0
10.1.2.0/24           Direct  0    0          10.1.2.1        GE0/0
10.1.2.1/32           Direct  0    0          127.0.0.1       InLoop0
10.1.3.0/24           Static  60   0          10.1.2.2        GE0/0
10.1.4.0/24           Static  60   0          10.1.2.2        GE0/0
127.0.0.0/8           Direct  0    0          127.0.0.1       InLoop0
127.0.0.1/32          Direct  0    0          127.0.0.1       InLoop0
```

六、实验思考

① 配置静态路由要注意哪些事项？

② 缺省路由适合用于什么场景？

实验十 RIP 协议配置

一、实验目的

① 熟悉 RIP 协议实现原理及报文格式；
② 了解 RIPv1 与 RIPv2 的区别；
③ 掌握 RIP 路由配置方法；
④ 掌握 RIP 动态路由信息的分析诊断方法。

二、实验内容

① 按图 10-1 连接实验设备，规划并配置接口地址；
② 分别启动配置 RIPv1、RIPv2 协议，实现设备互通；
③ 学习使用 Debug ip packet 和 Debug ip rip 命令，并对 debug 信息做分析；
④ 通过实验观察 RIPv1、RIPv2 的区别，给出分析过程与结果,实验 IP 采用 10.10.x.0 A 类私有网段；
⑤ 观察试验拓扑中链路状态发生改变时，路由表的前后信息对比及 debug 信息的变化。

三、实验设备及组网图

1. 实验设备

两台 H3C MSR30-20 路由器，一条专用 Console 配置线缆，两台计算机，超级终端或 Secure CRT 软件。

2. 组网图

RIP 协议配置实验组网如图 10-1 所示。

四、实验相关知识

1. RIP 简介

路由信息协议 RIP 是由施乐公司在 20 世纪 70 年代开发，专门为小型互联网而设计的一种较为简单的内部网关协议。在 1988 年第一版 RIPv1 被因特网协议组正式标准化为 RFC1058。1993 年又推出 RIP 协议的第二版 RIPv2，1994 年被标准化为 RFC1723。

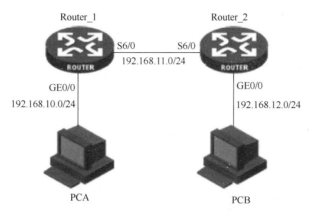

图 10-1　实验组网

RIP 是基于距离向量算法的路由协议，它使用“跳数”，即 metric 来衡量到达目标地址的路由距离。RIP 协议工作在 UDP 协议之上，它通过向相邻路由器广播 UDP 报文的交换路由信息，使用的 UDP 端口号是 520，如图 10-2 所示。在缺省情况下，RIP 路由器的路由更新时间是 30s，路由无效的时间是 180s，路由删除的时间是 300s。更新计时器用于记录时间。路由更新时，RIP 节点就会产生一系列包含自身全部路由表的报文。这些报文广播到每一个相邻节点。因此，每一个 RIP 路由器大约每隔 30s 应收到相邻 RIP 节点发来的路由信息用来维护自身路由表。

		BGP 179	RIP 520
IGRP 88	OSPF 89	TCP 6	UDP 17
IP			
CSMA/CD	TOKEN Ring	PPP	FR
物理接口			

图 10-2　路由协议栈

（1）RIPv1 报文格式

RIPv1 报文由头部和多个路由表项部分组成，一个 RIP 表项中最多可以有 25 个路由表项，如图 10-3 所示。RIP 是基于 UDP 协议的，所以 RIP 报文的数据包不能超过 512 个字节。RIPv1 没有掩码，不能使用可变长子网掩码和路由聚合特征，因此不能分割地址空间以最大效率应用有限的 IP 地址。

4字节		
命令	版本	必须为0
地址族标识		必须为0
IP地址		
必须为0		
必须为0		
距度(1～16)		

路由信息（20字节/路由记录）最多可重复25次

图 10-3　RIPv1 报文格式

命令：长度 8 位，报文类型分 request 报文和 reponse 报文，request 报文负责向邻居请求全部或者部分路由信息，reponse 报文发送自己全部或部分路由信息。

版本：长度 8 位，标识 RIP 的版本号。

必须为 0：长度 16 位，规定必须为零字段。

地址簇标识（address family identifier，AFI）：长度 16 位，其值为 2 时表示 IP 协议。

IP 地址：长度 32 位，该路由的目的 IP 地址，只能是自然网段的地址。

距离：长度 32 位，路由的开销值，即跳值。最大值 16 表示无穷大。

（2）RIPv2 报文格式

RIPv2 报文格式如图 10-4 所示。RIPv2 是一种无类别路由协议，与 RIPv1 相比，RIPv2 协议报文中携带掩码信息，支持无类域间路由（CIDR）、变长子网掩码（VLSMs）、验证、密钥管理、路由聚合；RIPv2 把和路由相关的子网掩码包含在路由更新报文中，实现了 VLSMs 和 CIDR，RIPv2 还增加了验证机制和报文组播等特性，组播地址为224.0.0.9。

图 10-4　RIPv2 报文格式

RIPv2 与 RIPv1 报文格式中不同的字段含义：

路由标记：路由标记字段的存在是为了支持外部网关协议。这个字段被期望用于传递自治系统的标号给外部网关协议及边界网关协议。

子网掩码：32 位，目的地址掩码。

下一跳路由地址：如果为 0.0.0.0，则表示发布此条路由信息的路由器地址就是最优下一跳地址，否则表示提供了一个比发布此条路由信息的路由器更优的下一条地址。

2．RIP 协议计时器

RIP 依赖计时器维护路由表，RIP 协议进程维护四种计时器。

更新计时器（Update Time）：RIP 协议平均每隔 30s 从每个启动 RIP 协议的接口不断发送响应消息。这个周期性的更新由更新计时器进行初始化，包含一个随机变化量用来防止表的同步。所以，实际更新时间为 25.5～30s，即 30s 减去一个在 4.5s 内的随机值。

超时计时器（Timeout Time）或无效计时器：当一条新的路由建立后，超时计时器就会被初始化，默认值为 180s，而每当接收到这条路由的更新报文时，超时计时器又将被重置成计时器的初始化值。如果一条路由的更新在 180s 内还没有收到，那么这条路由的跳数将变成 16，也就是标记为不可达路由，但不删除路由。

抑制计时器（Suppress Time）：如果路由器在相同的接口上收到一条路由更新的跳数大于路由选择表已记录的该路由的跳数，那么将启动一个抑制计时器，在抑制计时器的时间内该路由目的不可到达。如果在抑制计时器超时后还接收到该消息，那么这时路由器就认为该消息是真的，将修改路由。

垃圾收集计时器（Garbage-collect Time）或刷新计时器：一条路由在路由表中被标记为不可达后启动一个垃圾收集计时器，如果垃圾收集计时器也超时，则该路由将被通告为不可到达的路由，同时从路由选择表中删除该路由。

五、实验过程

1. 实验任务一：配置 RIPv1

（1）搭建实验环境，初始化实验设备

按图 10-1 所示搭建实验环境，清除配置文件，以出厂配置重启网络设备。

```
<H3C>reset saved-configuration
<H3C>reboot
……
<H3C>
```

（2）按表 10-1 所示更改设备名称并配置 PC、路由器端口 IP 地址

表 10-1　IP 地址规划表

设备名称	接口	IP 地址	网关
Router_1	S6/0	192.168.11.1/24	
	GE0/0	192.168.10.1/24	
Router_2	S6/0	192.168.11.2/24	
	GE0/0	192.168.12.1/24	
PCA		192.168.10.2/24	192.168.10.1/24
PCB		192.168.12.2/24	192.168.12.1/24

测试 PC 到网关的可达性：

```
C:\Documents and Settings\Administrator>ping 192.168.10.1
  Pinging 192.168.10.1 with 32 bytes of data:
    Reply from 192.168.10.1: bytes=32 time<1ms TTL=255
    Reply from 192.168.10.1: bytes=32 time<1ms TTL=255
    Reply from 192.168.10.1: bytes=32 time<1ms TTL=255
    Reply from 192.168.10.1: bytes=32 time<1ms TTL=255

  Ping statistics for 192.168.10.1:
      Packets: Sent = 4, Received = 4, Lost = 0 (0% loss),
  Approximate round trip times in milli-seconds:
      Minimum = 0ms, Maximum = 0ms, Average = 0ms
```

再测试 PC 之间的可达性。例如：在 PCA 上用 Ping 命令测试到 PCB 的可达性：

```
C:\Documents and Settings\Administrator>ping 192.168.12.2
  Pinging 192.168.12.2 with 32 bytes of data:
   Request timed out.
   Request timed out.
   Request timed out.
   Request timed out.

   Ping statistics for 192.168.12.2:
   Packets: Sent = 4, Received = 0, Lost = 4 (100% loss)
```

[Router_1]**display ip routing-table**　　//查看 Router_1 的路由表。
Routing Tables: Public
　　　　Destinations : 5　　　　Routes : 5
Destination/Mask　　　Proto　Pre　Cost　　　　NextHop　　　　Interface
127.0.0.0/8　　　　　Direct　0　　0　　　　　127.0.0.1　　　InLoop0
127.0.0.1/32　　　　 Direct　0　　0　　　　　127.0.0.1　　　InLoop0
192.168.11.0/24　　　Direct　0　　0　　　　　192.168.11.1　 S6/0
192.168.11.1/32　　　Direct　0　　0　　　　　127.0.0.1　　　InLoop0
192.168.11.2/32　　　Direct　0　　0　　　　　192.168.11.2

（3）启用 RIP

配置 Router_1：

[Router_1]**rip**
[Router_1-rip-1]**network 192.168.10.0**
[Router_1-rip-1]**network 192.168.11.0**

配置 Router_2：

[Router_2]**rip**
[Router_2-rip-1]**network 192.168.11.0**
[Router_2-rip-1]**network 192.168.12.0**

（4）查看测试分析

[Router_1]**display ip routing-table**　　　　　　　//查看 Router_1 的路由表。
Routing Tables: Public
　　　　Destinations : 6　　　　Routes : 6
Destination/Mask　　　Proto　Pre　Cost　　　　NextHop　　　　Interface
127.0.0.0/8　　　　　Direct　0　　0　　　　　127.0.0.1　　　InLoop0
127.0.0.1/32　　　　 Direct　0　　0　　　　　127.0.0.1　　　InLoop0
192.168.11.0/24　　　Direct　0　　0　　　　　192.168.11.1　 Eth0/0
192.168.11.1/32　　　Direct　0　　0　　　　　127.0.0.1　　　InLoop0
192.168.11.2/32　　　Direct　0　　0　　　　　192.168.11.2　 S6/0
192.168.12.0/24　　　RIP　　100　1　　　　　192.168.11.2

```
C:\Documents and Settings\Administrator>ping 192.168.12.2   //测试可达性。
  Pinging 192.168.12.2 with 32 bytes of data:
   Reply from 192.168.12.2: bytes=32 time=22ms TTL=62
```

```
    Reply from 192.168.12.2: bytes=32 time=21ms TTL=62
    Reply from 192.168.12.2: bytes=32 time=20ms TTL=62
    Reply from 192.168.12.2: bytes=32 time=20ms TTL=62

Ping statistics for 192.168.12.2:
    Packets: Sent = 4, Received = 4, Lost = 0 (0% loss),
Approximate round trip times in milli-seconds:
    Minimum = 20ms, Maximum = 22ms, Average = 20ms
```

可以看到 PC 之间实现了互通。

```
[Router_1]display rip                       //查看RIP运行状态。
Public VPN-instance name :
RIP process : 1
   RIP version : 1                          //RIP版本1。
   Preference : 100
   Checkzero : Enabled
   Default-cost : 0
   Summary : Enabled
   Hostroutes : Enabled
   Maximum number of balanced paths : 8
   Update time  :  30 sec(s)  Timeout time: 180 sec(s) //四个更新计时器。
   Suppress time : 120 sec(s)  Garbage-collect time :120 sec(s)
   update output delay :  20(ms)  output count :  3
   TRIP retransmit time :   5 sec(s)
   TRIP response packets retransmit count :  36
   Silent interfaces : None
   Default routes : Disabled
   Verify-source : Enabled
   Networks :
      192.168.11.0         192.168.10.0
   Configured peers : None
   Triggered updates sent : 1
   Number of routes changes : 1
   Number of replies to queries : 1
```

(5) 水平分割

在路由器启用了 RIP 之后，默认水平分割是开启的。以 Router_1 的 S6/0 接口为例，查看开启了水平分割之后路由的更新报文：

```
<Router_1>terminal monitor
<Router_1>terminal debugging
<Router_1>debugging rip 1 packet
*Apr 2 14:29:25:825 2010 Router_1 RM/6/RMDEBUG: RIP 1:Sending response on interface
Ethernet0/0 from 192.168.10.1to 255.255.255.255//RIPv1   //广播发送更新报文。
 *Apr  2  14:29:25:825  2010  Router_1 RM/6/RMDEBUG:    Packet : vers 1, cmd
response, length 44
```

*Apr　　2 14:29:25:826 2010 Router_1 RM/6/RMDEBUG:　　**AFI 2, dest 192.168.11.0, cost 1**
　　*Apr　　2 14:29:25:826 2010 Router_1 RM/6/RMDEBUG:　　**AFI 2, dest 192.168.12.0**, cost 2
　　*Apr　　2 14:29:25:826 2010 Router_1 RM/6/RMDEBUG: RIP 1 : **Sending** response on interface Serial1/0 from **192.168.11.1** to 255.255.255.255
　　*Apr　　2 14:29:25:826 2010 Router_1 RM/6/RMDEBUG:　　Packet : **vers 1**, cmd response, length 24
　　*Apr　　2 14:29:25:827 2010 Router_1 RM/6/RMDEBUG:　　AFI 2, **dest 192.168.10.0, cost 1**
　　*Apr　　2 14:29:39:725 2010 Router_1 RM/6/RMDEBUG: RIP 1 : Receive response from 192.168.11.2 on Serial1/0
　　*Apr　　2 14:29:39:725 2010 Router_1 RM/6/RMDEBUG:　　Packet : vers 1, cmd response, length 24
　　*Apr　　2 14:29:39:726 2010 Router_1 RM/6/RMDEBUG:　　AFI 2, dest 192.168.12.0, cost 1

　　观察在 Router_1 上的 S6/0 口上取消水平分割后，协议报文收发情况。

　　[Router_1-Serial1/0]**undo rip split-horizon**　　　　　　// 关闭水平分割
　　<Router_1>**debugging rip 1 packet**
　　*Apr2 14:38:49:825 2010 Router_1 RM/6/RMDEBUG: RIP 1 : Sending response on interface Ethernet0/0 from 192.168.10.1 to 255.255.255.255
　　*Apr2 14:38:49:825 2010 Router_1 RM/6/RMDEBUG: Packet : vers 1, cmd response, length 44
　　*Apr　2 14:38:49:826 2010 Router_1 RM/6/RMDEBUG:AFI 2, dest 192.168.11.0, cost 1
　　*Apr2 14:38:49:826 2010 Router_1 RM/6/RMDEBUG: AFI 2, dest 192.168.12.0, cost 2
　　*Apr2 14:38:49:826 2010 Router_1 RM/6/RMDEBUG: RIP 1 : **Sending** response on interface **Serial1/0** from **192.168.11.1** to 255.255.255.255
　　*Apr2 14:38:49:827 2010 Router_1 RM/6/RMDEBUG: Packet : vers 1, cmd response, length 64
　　*Apr2 14:38:49:827 2010 Router_1 RM/6/RMDEBUG: AFI 2, **dest 192.168.10.0, cost 1**
　　*Apr 2 14:38:49:827 2010 Router_1 RM/6/RMDEBUG: AFI 2, **dest 192.168.11.0, cost 1**
　　*Apr 2 14:38:49:827 2010 Router_1 RM/6/RMDEBUG: AFI 2, **dest 192.168.12.0, cost 2**
　　//Router_1 从 S6/0 接口发送了从该接口收到的路由 192.168.11.0 和 dest 192.168.12.0。
　　*Apr2 14:38:58:866 2010 Router_1 RM/6/RMDEBUG: RIP 1 : Receive response from 192.168.11.2 on Serial1/0
　　*Apr2 14:38:58:867 2010 Router_1 RM/6/RMDEBUG:　　Packet : vers 1, cmd response, length 24
　　*Apr2 14:38:58:867 2010 Router_1 RM/6/RMDEBUG:　　AFI 2, dest 192.168.12.0, cost

由以上输出比较发现，在水平分割功能关闭的情况下，Router_1 在接口 S6/0 上发送的路由包括从该接口收到的路由，而在启用水平分割的情况下，Router_1 是不会将从此接口收到的路由在更新协议包中进行发送。

（6）毒性逆转

另外一种避免环路的机制是毒性逆转。在 Router_1 的接口 S6/0 上启用毒性逆转，再观察收发协议报文的情况。

[Router_1-Serial1/0]**rip poison-reverse**　　　　//启用毒性逆转。
<Router_1>debugging rip 1 packet
　*Apr2 14:56:18:825 2010 Router_1 RM/6/RMDEBUG: RIP 1 : Sending response on interface Ethernet0/0 from 192.168.10.1 to 255.255.255.255
　*Apr2 14:56:18:825 2010 Router_1 RM/6/RMDEBUG:　Packet : vers 1, cmd response, length 44
　*Apr2 14:56:18:825 2010 Router_1 RM/6/RMDEBUG:　AFI 2, dest 192.168.11.0, cost 1
　*Apr2 14:56:18:826 2010 Router_1 RM/6/RMDEBUG: AFI 2, dest 192.168.12.0, cost 2
　*Apr2 14:56:18:826 2010 Router_1 RM/6/RMDEBUG: RIP 1 : **Sending** response on interface **Serial1/0** from 192.168.11.1 to 255.255.255.255
　*Apr2 14:56:18:826 2010 Router_1 RM/6/RMDEBUG: Packet : vers 1, cmd response, length 44
　*Apr2 14:56:18:827 2010 Router_1 RM/6/RMDEBUG: AFI 2, dest 192.168.10.0, cost 1
　*Apr2 14:56:18:827 2010 Router_1 RM/6/RMDEBUG:　AFI 2, **dest 192.168.12.0, cost 16**
　　　//Router_1 将从接口 S6/0 收到的路由度量值设置为无穷大发送出去。
　*Apr2 14:56:31:138 2010 Router_1 RM/6/RMDEBUG: RIP 1 : Receive response from 192.168.11.2 on Serial1/0
　*Apr2 14:56:31:138 2010 Router_1 RM/6/RMDEBUG: Packet : vers 1, cmd response, length 24
　*Apr2 14:56:31:138 2010 Router_1 RM/6/RMDEBUG:　AFI 2, dest 192.168.12.0, cost 1

由以上输出信息可知，启用毒性逆转后，Router_1 在接口 S6/0 上发送的路由更新包含了 192.168.12.0,但度量值为 16（无穷大）。Router_1 主动告诉 Router_2，从 Router_1 的 S6/0 口不能到达 192.168.12.0 网段。

（7）用 silent-interface 控制协议报文发送

在之前的试验中，路由器从每个启用 RIP 的接口收发协议报文，包括与 PC 相连的接口。事实上，PC 并不需要接收 RIP 协议报文，可以用 silent-interface 命令使接口只接收而不发送协议报文。

配置 Router_1：

[Router_1-rip-1]**silent-interface S0/0**

配置 Router_2：

[Router_2-rip-1]**silent-interface S0/0**

配置完成后，用 debugging 命令查看，可以发现，RIP 不再从接口 S0/0 发送协议报文。

RIP 是通过计时器来维护路由表的，路由器在 180s 后没收到对某路由项的更新，就会将该路由项从路由表中撤销。下面进行验证：

在 Router_2 的 S6/0 接口上使用 silent-interface 命令，在 Router_1 上查看路由表：

```
[Router_1]display ip routing-table
Routing Tables: Public
        Destinations : 6      Routes : 6

Destination/Mask       Proto   Pre Cost     NextHop           Interface

127.0.0.0/8            Direct  0   0        127.0.0.1         InLoop0
127.0.0.1/32           Direct  0   0        127.0.0.1         InLoop0
192.168.11.0/24        Direct  0   0        192.168.11.1      S6/0
192.168.11.1/32        Direct  0   0        127.0.0.1         InLoop0
192.168.11.2/32        Direct  0   0        192.168.11.2      S6/0
192.168.12.0/24        RIP     100 1        192.168.11.2      S6/0
```

可以看到，在 Router_1 的路由表中还有到 192.168.12.0 的路由。

180s 后，再来查看路由表：

```
[Router_1]dis ip routing-table
Routing Tables: Public
        Destinations : 7      Routes : 7

Destination/Mask       Proto   Pre Cost     NextHop           Interface

127.0.0.0/8            Direct  0   0        127.0.0.1         InLoop0
127.0.0.1/32           Direct  0   0        127.0.0.1         InLoop0
192.168.10.0/24        Direct  0   0        192.168.10.1      Eth0/0
192.168.10.1/32        Direct  0   0        127.0.0.1         InLoop0
192.168.11.0/24        Direct  0   0        192.168.11.1      S6/0
192.168.11.1/32        Direct  0   0        127.0.0.1         InLoop0
192.168.11.2/32        Direct  0   0        192.168.11.2      S6/0
```

可以看到路由 192.168.12.0 已经不存在了，查看 RIP 路由表：

```
[Router_1]dis rip 1 route
 Route Flags: R - RIP, T - TRIP
        P - Permanent, A - Aging, S - Suppressed, G - Garbage-collect
 ----------------------------------------------------------------------
 Peer 192.168.11.2  on Serial1/0
     Destination/Mask      Nexthop         Cost    Tag     Flags    Sec
     192.168.12.0/24       192.168.11.2    16      0       RSG      2
```

Garbage-collect 计时器发现在 RIP 路由表中还存在该路由，120s 以后再查看：

```
[Router_1]dis rip 1 route
 Route Flags: R - RIP, T - TRIP
```

```
             P - Permanent, A - Aging, S - Suppressed, G - Garbage-collect
```
此时,该路由项已经从路由器中被彻底删除了。

2. 实验任务二:RIPv2 的配置

RIPv1 协议本身存在一定的不足,在主类网络隔离非连续子网的情况下,不能够正确地接收到路由。RIPv2 弥补了 RIPv1 的不足,同时可以支持验证。

(1)搭建实验环境,初始化实验设备

按图 10-1 所示搭建实验环境,清除配置文件,以出厂配置重启网络设备。

```
<H3C>reset saved-configuration
<H3C>reboot
......
<H3C>
```

(2)按表 10-2 所示更改设备名称并配置 PC、路由器端口 IP 地址

表 10-2 IP 地址列表

设备名称	接口	IP 地址	网关
Router_1	S6/0	192.168.11.1/24	
	GE0/0	10.0.1.1/24	
Router_2	S6/0	192.168.11.2/24	
	GE0/0	10.0.0.1/24	
PCA		10.0.1.2/24	10.0.1.1/24
PCB		10.0.0.2/24	10.0.0.1/24

(3)配置 RIPv1,观察路由表

配置 Router_1:

```
[Router_1]rip
[Router_1-rip-1]network 192.168.11.0
[Router_1-rip-1]network 10.0.0.0
```

配置 Router_2:

```
[Router_2]rip
[Router_2-rip-1]network 192.168.11.0
[Router_2-rip-1]network 10.0.0.0

<Router_1>display ip routing-table          //查看路由表。
Routing Tables: Public
        Destinations : 7        Routes : 7

Destination/Mask    Proto  Pre  Cost     NextHop         Interface
10.0.1.0/24         Direct 0    0        10.0.1.1        Eth0/0
10.0.1.1/32         Direct 0    0        127.0.0.1       InLoop0
```

127.0.0.0/8	Direct	0	0	127.0.0.1	InLoop0
127.0.0.1/32	Direct	0	0	127.0.0.1	InLoop0
192.168.11.0/24	Direct	0	0	192.168.11.1	S6/0
192.168.11.1/32	Direct	0	0	127.0.0.1	InLoop0
192.168.11.2/32	Direct	0	0	192.168.11.2	S6/0

Router_1 的路由表中没有 10.0.0.0/24 网段的路由。

<Router_1>**debugging rip 1 packet**   //在 Router_1 上打开 debugging。

  Apr   2 16:15:54:825 2010 Router_1 RM/6/RMDEBUG: RIP 1 : Sending response on interface Ethernet0/0 from 10.0.1.1 to 255.255.255.255

  *Apr   2 16:15:54:825 2010 Router_1 RM/6/RMDEBUG: Packet : vers 1, cmd response, length 24

  *Apr   2 16:15:54:825 2010 Router_1 RM/6/RMDEBUG: AFI 2, dest 192.168.11.0, cost 1

  *Apr   2 16:15:54:826 2010 Router_1 RM/6/RMDEBUG: RIP 1 : Sending response on interface Serial1/0 from 192.168.11.1 to 255.255.255.255

  *Apr   2 16:15:54:826 2010 Router_1 RM/6/RMDEBUG: Packet : vers 1, cmd response, length 24

  *Apr   2 16:15:54:826 2010 Router_1 RM/6/RMDEBUG: AFI 2, dest 10.0.0.0, cost 1

  *Apr   2 16:16:05:356 2010 Router_1 RM/6/RMDEBUG: RIP 1 : **Receive** response from 192.168.11.2 on **Serial1/0**

  *Apr   2 16:16:05:356 2010 Router_1 RM/6/RMDEBUG: Packet : vers 1, cmd response, length 24

  *Apr   2 16:16:05:356 2010 Router_1 RM/6/RMDEBUG: AFI 2, dest 10.0.0.0, cost 1

  *Apr   2 16:16:05:357 2010 Router_1 RM/3/RMDEBUG: RIP 1 : **Ignoring route 10.0.0.0. Its major net addr is same as the local interface's.**

（4）配置 RIPv2

配置 Router_1：

[Router_1]**rip**
[Router_1-rip-1]**version 2**
[Router_1-rip-1]**undo summary**

配置 Router_2：

[Router_2]**rip**
[Router_2-rip-1]**version 2**
[Router_2-rip-1]**undo summary**

配置完成后，在 Router_1 上查看路由表，如下：

[Router_1]**display ip routing-table**
Routing Tables: Public
        Destinations : 6   Routes : 6

Destination/Mask	Proto	Pre	Cost	NextHop	Interface
10.0.0.0/24	RIP	100	1	192.168.11.2	S6/0
127.0.0.0/8	Direct	0	0	127.0.0.1	InLoop0
127.0.0.1/32	Direct	0	0	127.0.0.1	InLoop0
192.168.11.0/24	Direct	0	0	192.168.11.1	S6/0
192.168.11.1/32	Direct	0	0	127.0.0.1	InLoop0
192.168.11.2/32	Direct	0	0	192.168.11.2	S6/0

<Router_1>**debugging rip 1 packet**   //在 Router_1 上打开 debugging。
*Apr2 16:26:25:825 2010 Router_1 RM/6/RMDEBUG: RIP 1 : Sending response on interface Ethernet0/0 from 10.0.1.1 to **224.0.0.9**   //RIPv2 以组播形式发送协议报文。
*Apr2 16:26:25:825 2010 Router_1 RM/6/RMDEBUG: Packet : **vers 2**, cmd response, length 44
*Apr2 16:26:25:826 2010 Router_1 RM/6/RMDEBUG: AFI 2, dest 10.0.0.0/**255.255.255.0**, nexthop 0.0.0.0, cost 2, tag 0   //协议报文包含掩码信息。
*Apr2 16:26:25:826 2010 Router_1 RM/6/RMDEBUG: AFI 2, dest 192.168.11.0/**255.255.255.0**, nexthop 0.0.0.0, cost 1, tag 0
*Apr2 16:26:25:826 2010 Router_1 RM/6/RMDEBUG: RIP 1 : Sending response on interface Serial1/0 from 192.168.11.1 to **224.0.0.9**
*Apr2 16:26:25:826 2010 Router_1 RM/6/RMDEBUG: Packet : **vers 2**, cmd response, length 44
*Apr2 16:26:25:827 2010 Router_1 RM/6/RMDEBUG: AFI 2, dest 10.0.1.0/**255.255.255.0**, nexthop 0.0.0.0, cost 1, tag 0
*Apr2 16:26:25:827 2010 Router_1 RM/6/RMDEBUG: AFI 2, dest 10.0.0.0/**255.255.255.0**, nexthop 0.0.0.0, cost 16, tag 0
*Apr2 16:26:38:517 2010 Router_1 RM/6/RMDEBUG: RIP 1 : **Receive** response from 192.168.11.2 on Serial1/0
*Apr2 16:26:38:517 2010 Router_1 RM/6/RMDEBUG: Packet : **vers** 2, cmd response, length 24
*Apr2 16:26:38:517 2010 Router_1 RM/6/RMDEBUG: AFI 2, dest 10.0.0.0/**255.255.255.0**, nexthop 0.0.0.0, cost 1, tag 0

可以看到 RIPv2 是以组播的形式发送协议报文的，组播地址为 224.0.0.9。以组播的形式进行更新，既减少了资源的浪费，又最大限度地减少了对其他网络设备的干扰。同时，RIP 协议报文中包含掩码信息，保证了信息交互的正确性。

六、实验思考

① 简述"路由超时计时器"的作用是什么？
② RIP 使用 UDP 有何优点？
③ 为什么在 RIPv2 中引入组播？

实验十一　OSPF 协议配置

一、实验目的

① 熟悉 OSPF 协议实现原理及报文格式；
② 理解 RouterID、AreaID 的意义及邻接关系建立过程；
③ 了解广播网络 DR、BDR 选举过程；
④ 掌握单区域、多区域 OSPF 基本配置；
⑤ 掌握 OSPF 动态路由信息的分析诊断方法。

二、实验内容

① 按组网图连接实验设备，规划并配置接口地址；
② 配置广播网单区域 OSPF 路由协议；
③ 配置点对点网单区域 OSPF 路由协议；
④ 配置多区域 OSPF 路由协议；
⑤ 查看分析诊断 OSPF 邻接状态、链路状态数据库、路由表等信息。

三、实验设备与组网图

1. 实验设备

三台 H3C 路由器，一台交换机，一条专用 Console 配置线缆，两台计算机，超级终端或 Secure CRT 软件。

2. 组网图

OSPF 协议配置实验组网如图 11-1～图 11-3 所示。

四、实验相关知识

1. OSPF 协议

开放最短路径优先协议 OSPF 是 IETF 的内部网关协议工作组特意为 IP 网络而开发的一种路由协议。最初的 OSPF 规范在 RFC1131 中发布，1991 年在 RFC1247 中又推出

OSPF 版本 2。目前，针对 IPv4 协议使用的是 RFC 2328。

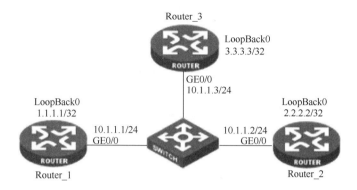

图 11-1 路由器之间通过交换机以太网相连

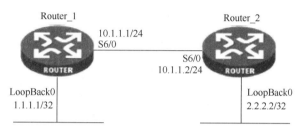

图 11-2 路由器之间用串口相连

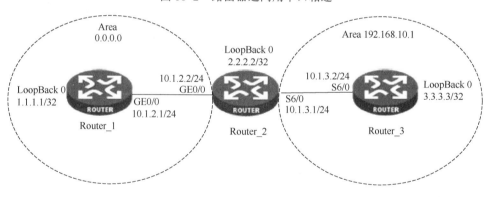

图 11-3 多区域 OSPF 实验组

OSPF 是基于链路状态 L-S 的路由协议。它通过传递链路状态通告得到网络状态信息，从而维护一张网络有向拓扑图-链路状态数据库，并利用最短路径优先算法（SPF 算法）获得路由域的最短路径优先树来维护 OSPF 路由表。如图 11-4 所示，OSPF 不用 UDP 而是直接用 IP 数据报传送，可见 OSPF 的位置在网络层，其协议号是 89，IP 服务类型为 0，优先级为网络控制级，并使用组播地址 224.0.0.5 来表示域中所有 SPF 路由器。

		BGP 179	RIP 520
IGRP 88	OSPF 89	TCP 6	UDP 17
IP			
CSMA/CD	TOKEN Ring	PPP	FR
物理接口			

图 11-4 路由协议栈

OSPF 路由表的变化基于网络中路由器物理连接的状态与速度变化，在缺省情况下，每30min 路由器会用泛洪法向所有路由器发送链路态通告信息。如有网络拓扑结构变化，链路状态会立即被广播到网络中的每一个路由器。如接口变化，信息立刻通过网络广播，如有多余路径，将重新计算 SPF 树。

为了使 OSPF 能够用于规模大的网络，OSPF 可将一个自治系统再划分为若干个更小的范围，叫作区域（Area）。每一个区域都有一个 32 位的区域标识符，用点分十进制表示，如 0.0.0.0。区域也不能太大，在一个区域内的路由器最好不超过 200 个。

一台路由器如果要运行 OSPF 协议，则必须存在路由器标识号（Router ID）。Router ID 是一个 32B 无符号整数，如 1.1.1.1，它可以在一个自治系统中唯一地标识一台路由器。Router ID 可以手工创建，也可以自动生成；如果没有通过命令指定 Router ID，将按照如下顺序自动生成一个 Router ID：

如果当前设备配置了 LoopBack 接口，将选取所有 LoopBack 接口上数值最大的 IP 地址作为 Router ID；

如果当前设备没有配置 LoopBack 接口，将选取它所有已经配置且链路有效的接口上数值最大的 IP 地址作为 Router ID。

2．OSPF 报文格式

OSPF 报文直接封装为 IP 报文协议，报文协议号为 89，如图 11-5 所示。

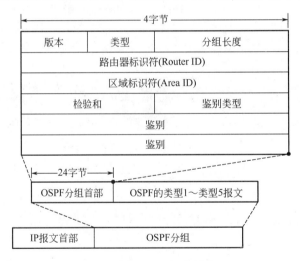

图 11-5　OSPF 报文格式

OSPF 分组首部包括以下内容。

版本：OSPF 的版本号。对于 OSPFv2 来说，其值为 2。

类型：OSPF 报文的类型。数值从 1 到 5，分别对应 Hello 报文、DD 报文、LSR 报文、LSU 报文和 LSAck 报文。

分组长度：OSPF 报文的总长度，包括报文头在内，单位为字节。

路由器标识符：始发该 LSA 的路由器的 ID。

区域标识符：始发 LSA 的路由器所在的区域 ID。

校验和：对整个报文的校验和。

鉴别类型：可分为不验证、明文口令验证和MD5验证，其值分别为0、1和2。

鉴别：其数值根据验证类型而定。当验证类型为0时，未作定义；为1时，此字段为密码信息；为2时，此字段包括Key ID、MD5验证数据长度和序列号的信息。MD5验证数据添加在OSPF报文后面，不包含在Authenticaiton字段中。

3. OSPF协议工作过程

OSPF协议工作过程主要有四个阶段：寻找邻居、建立邻接关系、链路状态信息传递实现链路状态数据库建立与维护、计算路由，如图11-6所示。

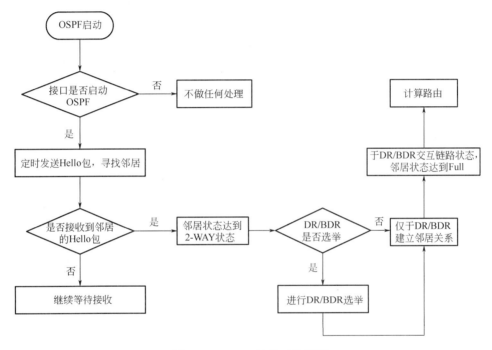

图11-6　OSPF协议工作流程

在OSPF中，邻居（neighbor）和邻接（adjacency）是两个不同的概念。如果两台路由器之间共享一条公共数据链路，OSPF路由器启动后，通过OSPF接口向外发送Hello报文。收到Hello报文的OSPF路由器会检查报文中所定义的参数，如果双方一致就会形成邻居关系。形成邻居关系的双方不一定都能形成邻接关系，这要根据网络类型而定。

只有当双方成功交换DD报文，交换LSA并达到LSDB的同步之后，才形成真正意义上的邻接关系。

（1）寻找邻居过程

寻找邻居过程如图11-7所示。

（2）建立邻接关系

建立邻接关系如图11-8、图11-9所示。

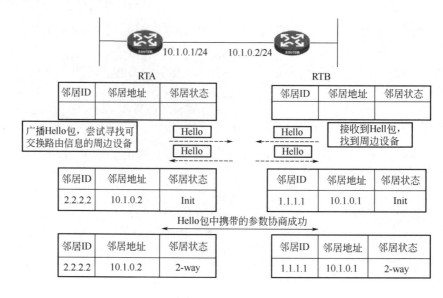

图 11-7 寻找邻居过程

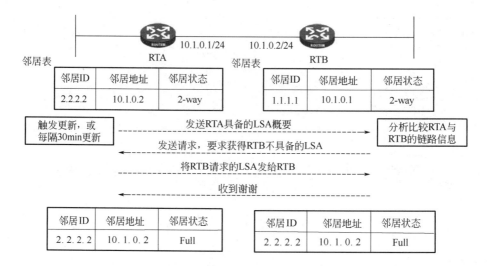

图 11-8 建立邻接关系（1）

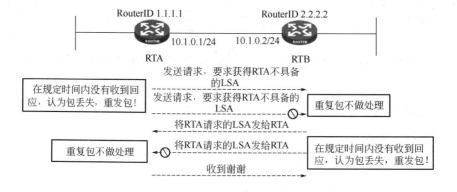

图 11-9 建立邻接关系（2）

（3）链路状态信息传递（链路状态数据库建立）

如图 11-10 所示，由于各路由器之间频繁地交换链路状态信息，因此所有的路由器最终都能建立一个链路状态数据库。这个数据库实际上就是全网的拓扑结构图，它在全网范围内是一致的，这称为链路状态数据库的同步。OSPF 的链路状态数据库能较快地进行更新，使各个路由器能及时更新其路由表。OSPF 的更新过程收敛得快是其重要优点。

（4）OSPF 协议计算路由

如图 11-10 所示，OSPF 协议路由的计算过程可简单描述如下：

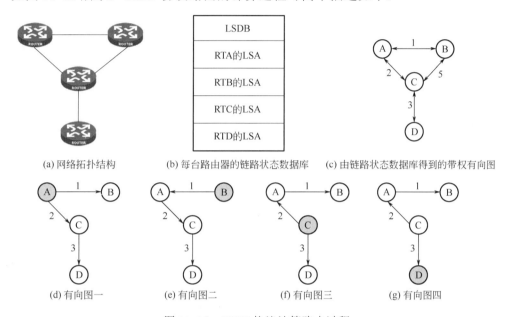

图 11-10　OSPF 协议计算路由过程

每台 OSPF 路由器根据周围的网络拓扑结构生成 LSA，并通过更新报文将 LSA 发送给网络中的其他 OSPF 路由器。

每台 OSPF 路由器都会收集其他路由器通告的 LSA，所有的 LSA 放在一起便组成了链路状态数据库（link state database，LSDB）。LSA 是对路由器周围网络拓扑结构的描述，LSDB 则是对整个自治系统的网络拓扑结构的描述。OSPF 路由器将 LSDB 转换成一张带权的有向图，这张图便是对整个网络拓扑结构的真实反映。各个路由器得到的有向图是完全相同的。

每台路由器根据有向图，使用 SPF 算法计算出一棵以自己为根的最短路径树，这棵树给出了到自治系统中各节点的路由。

五、实验过程

1. 实验任务一：广播网单区域 OSPF 配置

（1）搭建实验环境，初始化实验设备

按图 11-1 所示搭建实验环境，清除配置文件，以出厂配置重启网络设备。

```
<H3C>reset saved-configuration
<H3C>reboot
......
<H3C>
```

（2）按表 11-1 所示更改设备名称并配置 PC、路由器端口 IP 地址

表 11-1　IP 地址规划表

设备名称	接口	IP 地址	网关
Router_1	GE0/0	10.1.1.1/24	
	LoopBack 0	1.1.1.1/32	
Router_2	GE0/0	10.1.1.2/24	
	LoopBack 0	2.2.2.2/32	
Router_3	GE0/0	10.1.1.3/24	
	LoopBack 0	3.3.3.3/32	
Switch			

（3）配置 OSPF

配置 Router_1：

```
[Router_1]interface GE0/0
[Router_1- GE0/0]ip add 10.1.1.1 24
[Router_1]interface loopback 0
[Router_1-LoopBack0]ip address 1.1.1.1 32
[Router_1]router id 1.1.1.1
[Router_1]ospf 1
[Router_1-ospf-1]area 0
[Router_2-ospf-1-area-0.0.0.0]network 1.1.1.1 0.0.0.0
[Router_2-ospf-1-area-0.0.0.0]network 10.1.1.0 0.0.0.255
```

配置 Router_2：

```
[Router_2]int GE0/0
[Router_2- GE0/0]ip add 10.1.1.2 24
[Router_2]interface loopback 0
[Router_2-LoopBack0]ip address 2.2.2.2 32
[Router_2]router id 2.2.2.2
[Router_2]ospf 1
[Router_2-ospf-1]area 0.0.0.0
[Router_2-ospf-1-area-0.0.0.0]network 2.2.2.2 0.0.0.0
[Router_2-ospf-1-area-0.0.0.0]network 10.1.1.0 0.0.0.255
```

配置 Router_3：

```
[Router_3]int GE0/0
[Router_3- GE0/0]ip add 10.1.1.3 24
[Router_3]interface loopback 0
[Router_3-LoopBack0]ip address 3.3.3.3 32
```

```
[Router_3]router id 3.3.3.3
[Router_3]ospf 1
[Router_3-ospf-1]area 0.0.0.0
[Router_3-ospf-1-area-0.0.0.0]network 3.3.3.3 0.0.0.0
[Router_3-ospf-1-area-0.0.0.0]network 10.1.1.0 0.0.0.255
```

(4）检查路由器 OSPF 邻居状态、链路状态数据库及路由表

1）查看路由器

分别查看路由器 Router_1、Router_2、Router_3 的 OSPF 邻居状态，显示如下：

```
[Router_1]display ospf peer
           OSPF Process 1 with Router ID 1.1.1.1
                Neighbor Brief Information


Area: 0.0.0.0
Router ID        Address       Pri  Dead-Time  Interface    State
2.2.2.2          10.1.1.2      1    39         GE0/0        Full/BDR
3.3.3.3          10.1.1.3      1    35         GE0/0        Full/DROther
```

2）显示路由器 Router_1 的 OSPF 的链路状态数据库

```
[Router_1]display ospf lsdb
      OSPF Process 1 with Router ID 1.1.1.1
              Link State Database


Area: 0.0.0.0
Type       LinkState ID      AdvRouter        Age    Len   Sequence    Metric
Router     3.3.3.3           3.3.3.3          110    48    80000003    0
Router     1.1.1.1           1.1.1.1          113    48    80000006    0
Router     2.2.2.2           2.2.2.2          114    48    80000005    0
Network    10.1.1.1          1.1.1.1          108    36    80000004    0
```

3）显示 OSPF 路由信息

```
[Router_1]display ospf routing
              OSPF Process 1 with Router ID 1.1.1.1
                                     Routing Tables
Routing for Network
Destination       Cost    Type     NextHop      AdvRouter    Area
3.3.3.3/32        1       Stub     10.1.1.3     3.3.3.3      0.0.0.0
2.2.2.2/32        1       Stub     10.1.1.2     2.2.2.2      0.0.0.0
10.1.1.0/24       1       Transit  10.1.1.1     1.1.1.1      0.0.0.0
1.1.1.1/32        0       Stub     1.1.1.1      1.1.1.1      0.0.0.0
Total Nets: 4
Intra Area: 4  Inter Area: 0  ASE: 0  NSSA: 0
```

4）查看全局路由表

在 Router_1 上查看全局路由表，显示如下：

```
[Router_1]display ip routing-table
Routing Tables: Public
        Destinations : 7         Routes : 7
Destination/Mask    Proto  Pre  Cost       NextHop         Interface
1.1.1.1/32          Direct 0    0          127.0.0.1       InLoop0
2.2.2.2/32          OSPF   10   1          10.1.1.2        GE0/0
3.3.3.3/32          OSPF   10   1          10.1.1.3        GE0/0
10.1.1.0/24         Direct 0    0          10.1.1.1        GE0/0
10.1.1.1/32         Direct 0    0          127.0.0.1       InLoop0
127.0.0.0/8         Direct 0    0          127.0.0.1       InLoop0
127.0.0.1/32        Direct 0    0          127.0.0.1       InLoop0
```

（5）测试网络连通性

在 Router_1 上，用 Ping 命令测试显示如下：

```
[Router_1]ping -a 1.1.1.1  2.2.2.2
  PING 2.2.2.2: 56  data bytes, press CTRL_C to break
    Reply from 2.2.2.2: bytes=56 Sequence=1 ttl=255 time=3 ms
    Reply from 2.2.2.2: bytes=56 Sequence=2 ttl=255 time=1 ms
    Reply from 2.2.2.2: bytes=56 Sequence=3 ttl=255 time=1 ms
    Reply from 2.2.2.2: bytes=56 Sequence=4 ttl=255 time=1 ms
    Reply from 2.2.2.2: bytes=56 Sequence=5 ttl=255 time=1 ms

 --- 2.2.2.2 ping statistics ---
  5 packet(s) tRouter_1nsmitted
  5 packet(s) received
  0.00% packet loss
  round-trip min/avg/max = 1/1/3 ms
```

2. 实验任务二：点对点网络单区域 OSPF 配置

（1）搭建实验环境，初始化实验设备

按图 11-2 所示路由器之间用串口线相连搭建实验环境，清除配置文件，以出厂配置重启路由器。

```
<H3C>reset saved-configuration
<H3C>reboot
……
<H3C>
```

（2）按表 11-2 所示更改设备名称并配置 PC、路由器端口 IP 地址

表 11-2 IP 地址规划表

设备名称	接口	IP 地址	网关
Router_1	S6/0	10.1.1.1/24	
	Loopback 0	1.1.1.1/32	
Router_2	S6/0	10.1.1.2/24	
	Loopback 0	2.2.2.2/32	

（3）配置 OSPF

配置 Router_1：

[Router_1]**interface s6/0**
[Router_1-Serial6/0]**ip add 10.1.1.1 24**
[Router_1]**interface loopback 0**
[Router_1-LoopBack0]**ip address 1.1.1.1 32**
[Router_1]**router id 1.1.1.1**
[Router_1]**ospf 1**
[Router_1-ospf-1]**area 0**
[Router_1-ospf-1-area-0.0.0.0]**network 1.1.1.1 0.0.0.0**
[Router_1-ospf-1-area-0.0.0.0]**network 10.1.1.0 0.0.0.255**

配置 Router_2：

[Router_2]**interface s6/0**
[Router_2-Serial6/0]**ip add 10.1.1.2 24**
[Router_2]**interface loopback 0**
[Router_2-LoopBack0]**ip address 2.2.2.2 32**
[Router_2]**router id 2.2.2.2**
[Router_2]**ospf 1**
[Router_2-ospf-1]**area 0.0.0.0**
[Router_2-ospf-1-area-0.0.0.0]**network 2.2.2.2 0.0.0.0**
[Router_2-ospf-1-area-0.0.0.0]**network 10.1.1.0 0.0.0.255**

（4）检查路由器 OSPF 邻居状态、链路状态数据库及路由表

1）查看路由器 Router_1、Router_2 的 OSPF 邻居状态

[Router_1]**display ospf peer**

```
          OSPF Process 1 with Router ID 1.1.1.1
                 Neighbor Brief Information

 Area: 0.0.0.0
 Router ID       Address        Pri  Dead-Time  Interface    State
 2.2.2.2         10.1.1.2       1    32         S6/0         Full/ -
```

[Router_2]**display ospf peer**

```
          OSPF Process 1 with Router ID 2.2.2.2
                 Neighbor Brief Information

 Area: 0.0.0.0
 Router ID       Address        Pri  Dead-Time  Interface    State
 1.1.1.1         10.1.1.1       1    32         S6/0         Full/ -
```

2）显示路由器 Router_1 的 OSPF 的链路状态数据库

[Router_1]**display ospf lsdb**

```
      OSPF Process 1 with Router ID 192.168.1.1
              Link State Database
                  Area: 0.0.0.0
```

Type	LinkState ID	AdvRouter	Age	Len	Sequence	Metric
Router	192.168.1.1	192.168.1.1	35	60	80000005	0
Router	2.2.2.2	2.2.2.2				

3）在 Router_1 上查看路由器的 OSPF 路由表

```
[Router_1]display ospf routing
         OSPF Process 1 with Router ID 1.1.1.1
                 Routing Tables

Routing for Network
Destination        Cost     Type      NextHop        AdvRouter        Area
2.2.2.2/32         1        Stub      10.1.1.2       2.2.2.2          0.0.0.0
10.1.1.0/24        1        TRouter_1nsit 10.1.1.1   1.1.1.1          0.0.0.0
1.1.1.1/32         0        Stub      1.1.1.1        1.1.1.1          0.0.0.0
Total Nets: 3
IntRouter_1 Area: 3  Inter Area: 0  ASE: 0  NSSA: 0
```

4）在 Router_1 上查看全局路由表

```
[Router_1]display ip routing-table
Routing Tables: Public
         Destinations : 7     Routes : 7

Destination/Mask        Proto    Pre    Cost     NextHop         Interface

1.1.1.1/32              Direct   0      0        127.0.0.1       InLoop0
2.2.2.2/32              OSPF     10     1562     10.1.1.2        S6/0
10.1.1.0/24             Direct   0      0        10.1.1.1        S6/0
10.1.1.1/32             Direct   0      0        127.0.0.1       InLoop0
10.1.1.2/32             Direct   0      0        10.1.1.2        S6/0
127.0.0.0/8             Direct   0      0        127.0.0.1       InLoop0
127.0.0.1/32            Direct   0      0        127.0.0.1       InLoop0
```

（5）测试网络连通性

在 Router_1 上用 Ping 命令测试显示如下：

```
[Router_1]ping -a 1.1.1.1 2.2.2.2
 PING 2.2.2.2: 56  data bytes, press CTRL_C to break
   Reply from 2.2.2.2: bytes=56 Sequence=1 ttl=255 time=3 ms
   Reply from 2.2.2.2: bytes=56 Sequence=2 ttl=255 time=1 ms
   Reply from 2.2.2.2: bytes=56 Sequence=3 ttl=255 time=1 ms
   Reply from 2.2.2.2: bytes=56 Sequence=4 ttl=255 time=1 ms
   Reply from 2.2.2.2: bytes=56 Sequence=5 ttl=255 time=1 ms

 --- 2.2.2.2 ping statistics ---
  5 packet(s) tRouter_1nsmitted
  5 packet(s) received
  0.00% packet loss
  round-trip min/avg/max = 1/1/3 ms
```

3. 实验任务三：多区域 OSPF 配置

（1）搭建实验环境

按图 11-3 所示搭建实验环境，清除配置文件，以出厂配置重启路由器。

<H3C>**reset saved-configuration**
<H3C>**reboot**
......
<H3C>

（2）按表 11-3 所示更改设备名称并配置 PC、路由器端口 IP 地址

表 11-3 IP 地址规划表

设备名称	接口	IP 地址	网关
Router_1	GE0/0	10.1.2.1/24	
	LoopBack 0	1.1.1.1/32	
Router_2	GE0/0	10.1.2.2/24	
	LoopBack 0	2.2.2.2/32	
	S6/0	10.1.3.1/24	
Router_3	S6/0	10.1.3.2/24	
	LoopBack 0	3.3.3.3/32	

（3）配置 OSPF

配置 Router_1：

[Router_1]**interface GE0/0**
[Router_1- GE0/0]**ip add 10.1.2.1 24**
[Router_1]**interface loopback 0**
[Router_1-LoopBack0]**ip address 1.1.1.1 32**
[Router_1]**router id 1.1.1.1**
[Router_1]**ospf 1**
[Router_1-ospf-1]**area 0**
[Router_2-ospf-1-area-0.0.0.0]**network 1.1.1.1 0.0.0.0**
[Router_2-ospf-1-area-0.0.0.0]**network 10.1.2.0 0.0.0.255**

配置 Router_2：

[Router_2]
[Router_2]**int GE0/0**
[Router_2- GE0/0]**ip add 10.1.2.2 24**
[Router_2]**interface loopback 0**
[Router_2-LoopBack0]**ip address 2.2.2.2 32**
[Router_2]**router id 2.2.2.2**
[Router_2]**ospf 1**

```
[Router_2-ospf-1]area 0.0.0.0
[Router_2-ospf-1-area-0.0.0.0]network 2.2.2.2 0.0.0.0
[Router_2-ospf-1-area-0.0.0.0]network 10.1.2.0 0.0.0.255
[Router_2]
[Router_2]int s6/0
[Router_2-Serial6/0]ip add 10.1.3.1 24
[Router_2]ospf 1
[Router_2-ospf-1]area 192.168.10.1
[Router_2-ospf-1-area-0.0.0.0]network 2.2.2.2 0.0.0.0
[Router_2-ospf-1-area-0.0.0.0]network 10.1.3.0 0.0.0.255
```

配置 Router_3：

```
[Router_3]
[Router_3]int S6/0
[Router_3- S6/0]ip add 10.1.3.2 24
[Router_3]interface loopback 0
[Router_3-LoopBack0]ip address 3.3.3.3 32
[Router_3]router id 3.3.3.3
[Router_3]ospf 1
[Router_3-ospf-1]area 192.168.10.1
[Router_3-ospf-1-area-192.168.10.1]network 3.3.3.3 0.0.0.0
[Router_3-ospf-1-area-192.168.10.1]network 10.1.3.0 0.0.0.255
```

（4）查看分析 OSPF 邻接状态、链路状态数据库、路由表

```
[Router_1]display ospf peer
 OSPF Process 1 with Router ID 1.1.1.1
                Neighbor Brief Information

 Area: 0.0.0.0
 Router ID       Address        Pri   Dead-Time    Interface      State
 2.2.2.2         10.1.2.2       1     39           GE0/0          Full/BDR
[Router_1]display ospf lsdb

[Router_1]display ospf routing
     OSPF Process 1 with Router ID 1.1.1.1
               Routing Tables
 Routing for Network
 Destination     Cost    Type       NextHop        AdvRouter      Area
 3.3.3.3/32      1563    Inter      10.1.2.2       2.2.2.2        0.0.0.0
 2.2.2.2/32      1       Stub       10.1.2.2       2.2.2.2        0.0.0.0
 1.1.1.1/32      0       Stub       1.1.1.1        1.1.1.1        0.0.0.0
 10.1.2.0/24     1       Transit    10.1.2.1       1.1.1.1        0.0.0.0
 10.1.3.0/24     1563    Inter      10.1.2.2       2.2.2.2        0.0.0.0
 Total Nets: 5
 Intra Area: 3  Inter Area: 2  ASE: 0  NSSA: 0

[Router_1]display ip routing-table
 Routing Tables: Public
         Destinations : 8       Routes : 8
```

```
Destination/Mask      Proto   Pre Cost        NextHop         Interface
1.1.1.1/32            Direct  0   0           127.0.0.1       InLoop0
2.2.2.2/32            OSPF    10  1           10.1.2.2        GE0/0
3.3.3.3/32            OSPF    10  1563        10.1.2.2        GE0/0
10.1.2.0/24           Direct  0   0           10.1.2.1        GE0/0
10.1.2.1/32           Direct  0   0           127.0.0.1       InLoop0
10.1.3.0/24           OSPF    10  1563        10.1.2.2        GE0/0
127.0.0.0/8           Direct  0   0           127.0.0.1       InLoop0
127.0.0.1/32          Direct  0   0           127.0.0.1       InLoop0
```

4．选做实验任务：通过修改接口优先级改变 DR 和 BDR 的选举结果

修改路由器接口优先级。

在 Router_2 的 GE0/0 上修改优先级为 0。

[Router_2]**interface GE 0/0**
[Router_2-GE0/0]**ospf dr-priority 0**

重启 OSPF 进程。

先将 Router_2 上的 OSPF 进程重启，再将 Router_1 的 OSPF 进程重启。

```
<Router_2>reset ospf 1 process
Warning : Reset OSPF process? [Y/N]:y
<Router_1>reset ospf 1 process
Warning : Reset OSPF process? [Y/N]:y
```

查看路由器的 OSPF 邻居状态：

[Router_1]**display ospf peer**

```
            OSPF Process 1 with Router ID 1.1.1.1
                 Neighbor Brief Information
Area: 0.0.0.0
Router ID       Address         Pri Dead-Time Interface      State
2.2.2.2         10.1.2.2        0   32        GE0/0          2-Way/ -
```

虽然 Router_2 先启动，但是由于 Router_2 的 E0/0 的优先级为 0，不具备选举资格，所以 Router_2 是 DROther。

六、实验思考

① 为什么在邻居列表中 Router ID 最小的路由器会被选为 DR？请解释它们选举 DR、BDR 的过程。

② Router_1 和 Router_2 优先级相同，并且 Router_2 的 Router ID 更大，为什么 Router_1 被选为 DR？

③ 为什么要首先选用逻辑接口地址作为路由器 ID？

实验十二　PPP 配置

一、实验目的

① 掌握 PPP 的基本配置；
② 掌握 PPP PAP 的配置；
③ 掌握 PPP CHAP 的配置。

二、实验内容

① 按图 12-1 连接实验设备；
② 设备接口封装 PPP 协议、配置 IP 地址，并查看相应接口信息；
③ 配置 PAP 认证的单向验证和双向验证，并进行相应的测试；
④ 配置 CHAP 认证，并进行相应的测试。

三、实验设备与组网图

1．实验设备

两台 H3C 路由器，一条专用 Console 配置线缆，一台计算机，超级终端或 Secure CRT 软件。

2．组网图

广域网协议需要适应多变的网络类型，HDLC 只支持同步串行链路，并且不支持验证。而 PPP 支持同/异步线路，能够提供验证，易于扩展。

实验组网如图 12-1 所示，使用两台 MSR 系列路由器，用串口相连。

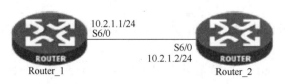

图 12-1　路由器之间用串口相连

四、实验相关知识

1. PPP 简介

PPP 协议是在点到点链路上承载网络层数据包的一种链路层协议，由于它能够提供用户验证、易于扩充，并且支持同/异步通信，因而获得广泛应用。

PPP 定义了一整的协议，包括链路控制协议、网络层控制协议和验证协议。

链路控制协议（link control protocol，LCP）：主要用来建立、拆除和监控数据链路。PPP 提供的 LCP 功能全面，适用于大多数环境。LCP 用于就封装格式选项自动达成一致，处理数据包大小限制，探测环路链路和其他普通的配置错误及终止链路。LCP 提供的其他可选功能有认证链路中对等单元的身份、决定链路功能正常或链路失败情况。

网络控制协议（network control protocol，NCP）：一种扩展链路控制协议，主要用来协商在该数据链路上所传输的数据包的格式与类型。用于建立、配置、测试和管理数据链路连接。

用于网络安全方面的验证协议族 PAP 和 CHAP。

为了建立点对点链路通信，PPP 链路的每一端必须首先发送 LCP 包，以便设定和测试数据链路。在链路建立，LCP 所需的可选功能被选定之后，PPP 必须发送 NCP 包，以便选择和设定一个或更多的网络层协议。一旦每个被选择的网络层协议都被设定好了，来自每个网络层协议的数据包就能在链路上发送了。

链路将保持通信设定不变，直到有 LCP 和 NCP 数据包关闭链路，或者发生一些外部事件，如休止状态的定时器期满或者网络管理员干涉。

2. PPP 链路建立过程

PPP 协议中提供了一整套方案来解决链路建立、维护、拆除、上层协议协商、认证等问题。PPP 协议包含这样几个部分：链路控制协议 LCP；网络控制协议 NCP；认证协议，最常用的包括口令验证协议 PAP（password authentication protocol）和挑战握手验证协议 CHAP（challenge-handshake authentication protocol）。

LCP 负责创建，维护或终止一次物理连接。NCP 是一族协议，负责解决物理连接上运行什么网络协议，以及解决上层网络协议发生的问题。

一个典型的链路建立过程分为三个阶段：创建阶段、认证阶段和网络协商阶段。

3. PPP 口令验证方式

（1）口令验证协议

PAP 是一种简单的明文验证方式。网络接入服务器（network access server，NAS）要求用户提供用户名和口令，PAP 以明文方式返回用户信息。很明显，这种验证方式的安全性较差，第三方可以很容易地获取被传送的用户名和口令，并利用这些信息与 NAS 建立连接，获取 NAS 提供的所有资源。所以，一旦用户密码被第三方窃取，PAP 无法提供避免受到第三方攻击的保障措施。

PAP 验证为两次握手验证，密码为明文，PAP 验证的过程如下：

① 被验证方发送用户名和密码到验证方；

② 验证方根据本端用户表查看是否有此用户及密码是否正确，然后返回不同的响应。

PAP 不是一种安全的验证协议。当验证时，口令以明文方式在链路上发送，并且由于完成 PPP 链路建立后，被验证方会不停地在链路上反复发送用户名和口令，直到身份验证过程结束，所以不能防止攻击。

（2）挑战-握手验证协议

CHAP 是一种加密的验证方式，能够避免建立连接时传送用户的真实密码。NAS 向远程用户发送一个挑战口令，其中包括会话 ID 和一个任意生成的挑战字串。远程客户必须使用 MD5 单向哈希算法，返回用户名和加密的挑战口令，会话 ID 及用户口令，其中用户名以非哈希方式发送。

CHAP 对 PAP 进行了改进，不再直接通过链路发送明文口令，而是使用挑战口令以哈希算法对口令进行加密。因为服务器端存有客户的明文口令，所以服务器可以重复客户端进行的操作，并将结果与用户返回的口令进行对照。CHAP 为每一次验证任意生成一个挑战字符串来防止受到重放攻击。在整个连接过程中，CHAP 将不定时向客户端重复发送挑战口令，从而避免第 3 方冒充远程客户进行攻击。

在某些连接时，在允许网络层协议数据包交换之前希望对对等实体进行认证。缺省时，认证不是必要的。如果应用时希望对等实体使用某些认证协议进行认证，这种要求必须在建立连接阶段提出。

认证阶段应该紧接在建立连接阶段后。然而，可能有连接质量的决定并行出现。应用时绝对不允许连接质量决定数据包的交换使认证有不确定的延迟。认证阶段后的网络层协议阶段必须等到认证结束后才能开始。如果认证失败，将转而进入终止连接阶段。仅仅是连接控制协议、认证协议、连接质量监测的数据包才被允许在此阶段中出现。所有其他在此阶段中接收到的数据包都将被静默丢弃。

如果对方拒绝认证，己方有权进入终止连接阶段。

CHAP 验证过程如下：

① 验证方主动发起验证请求，验证方向被验证方发送一些随机产生的报文，并同时附带本端的用户名一起发送给被验证方；

② 被验证方接到验证方的验证请求后，检查本端接口上是否配置了缺省的 CHAP 码，如果配置了，则被验证方利用报文 ID、该缺省密码和 MD5 算法对该随机报文进行加密，将生成的密文和自己的用户名发回验证方；

③ 如果被验证方检查发现本端接口上没有配置缺省的 CHAP 密码，则被验证方根据此报文中验证方的用户名在本端的用户表查找该用户对应的密码，如果在用户表找到了与验证方用户名相同的用户，便利用报文 ID、此用户的密钥（密码）和 MD5 算法对该随机报文进行加密，将生成的密文和被验证方自己的用户名发回验证方；

④ 验证方用自己保存的被验证方密码和 MD5 算法对原随机报文加密，比较二者的密文，根据比较结果返回不同的响应：Acknowledge 或 Not Acknowledge。

五、实验过程

1. 实验任务一：PPP 协议基本配置

（1）搭建实验环境及基本配置

按图 12-1 所示搭建实验环境，清除配置文件，以出厂配置重启路由器。

```
<H3C>reset saved-configuration
<H3C>reboot
……
```

（2）给 Router_1 上 S6/0 接口封装 PPP 协议、配置好 IP 地址，并查看相应接口信息
Router_1:

```
[Router_1]interface Serial 6/0
[Router_1-Serial6/0]link-protocol ppp
[Router_1-Serial6/0]ip address 10.2.1.1 24
```

通过 display interface 命令查看接口信息：

```
[Router_1]display interface Serial 6/0
Serial6/0 current state :UP                         //物理接口 UP。
Line protocol current state :UP                     // PPP 协议 UP。
Description : Serial6/0 Interface
The Maximum Transmit Unit is 1500, Hold timer is 1(sec)
Internet Address is 10.2.1.1/24
Link layer protocol is PPP                          //链路类型为 PPP。
LCP opened, IPCP opened, OSICP opened               //LCP 和 IPCP 状态为 opened。
Output queue :（Urgent queuing : Size/Length/Discards） 0/51/0
Output queue :（Protocol queuing : Size/Length/Discards）0/501/0
Output queue :（FIFO queuing : Size/Length/Discards） 0/75/0
Physical layer is synchronous,
Interface is DTE, Cable type is V24
Last clearing of counters: Never
Last 300 seconds input rate 14.02 bytes/sec, 112 位s/sec, 0.85 packets/ sec
Last 300 seconds output rate 21.83 bytes/sec, 174 位s/sec, 1.21 packets/ sec
Input: 414 packets, 7706 bytes
0 broadcasts, 0 multicasts
0 errors, 0 runts, 0 giants
0 CRC, 0 align errors, 0 overruns
0 dribbles, 0 aborts, 0 no buffers
0 frame errors
Output:623 packets, 13616 bytes
0 errors, 0 underruns, 0 collisions
0 deferred
DCD=UP  DTR=UP  DSR=UP  RTS=UP  CTS=UP
```

（3）给 Router_2 上 S6/0 接口封装 PPP 协议、配置好 IP 地址，并查看相应接口信息
配置 Router_2:

```
[Router_2]interface Serial 6/0
[Router_2-Serial6/0]link-protocol ppp
[Router_2-Serial6/0]ip address 10.2.1.2 24
```

通过 display interface 命令查看接口信息。

```
[Router_2]display interface Serial 6/0
Serial6/0 current state :UP
Line protocol current state :UP
Description : Serial6/0 Interface
The Maximum Transmit Unit is 1500, Hold timer is 1(sec)
Internet Address is 10.2.1.2/24
Link layer protocol is PPP
LCP opened, IPCP opened
Output queue : (Urgent queuing : Size/Length/Discards)  0/51/0
Output queue : (Protocol queuing : Size/Length/Discards)  0/501/0
Output queue : (FIFO queuing : Size/Length/Discards)  0/75/0
Physical layer is synchronous,Baudrate is 64000 bps
Interface is DCE, Cable type is V24
Last clearing of counters: Never
Last 300 seconds input rate 21.63 bytes/sec, 173 位s/sec, 1.13 packets/sec
Last 300 seconds output rate 12.85 bytes/sec, 102 位s/sec, 0.74 packets/sec
Input: 814 packets, 15882 bytes
3 broadcasts, 0 multicasts
0 errors, 0 runts, 0 giants
0 CRC, 0 align errors, 0 overruns
0 dribbles, 0 aborts, 0 no buffers
0 frame errors
Output:618 packets, 10234 bytes
0 errors, 0 underruns, 0 collisions
0 deferred
DCD=UP  DTR=UP  DSR=UP  RTS=UP  CTS=UP
```

（4）检查两台 Router_1 与 Router_2 的连通性

在 Router_1 上 Ping Router_2 的接口地址，显示如下。

配置 Router_1：

```
[Router_1]ping 10.2.1.2
PING 10.2.1.2: 56  data bytes, press CTRL_C to break
Reply from 10.2.1.2: bytes=56 Sequence=1 ttl=255 time=27 ms
Reply from 10.2.1.2: bytes=56 Sequence=2 ttl=255 time=27 ms
Reply from 10.2.1.2: bytes=56 Sequence=3 ttl=255 time=27 ms
Reply from 10.2.1.2: bytes=56 Sequence=4 ttl=255 time=27 ms
Reply from 10.2.1.2: bytes=56 Sequence=5 ttl=255 time=27 ms

--- 10.2.1.2 ping statistics ---
5 packet(s) transmitted
5 packet(s) received
0.00% packet loss
round-trip min/avg/max = 27/27/27 ms
```

2. 实验任务二：PPP PAP 认证配置

实验要求 Router_1 和 Router_2 之间采用 PAP 验证，Router_1 作为主验证方。

（1）在 Router_1 上配置本地以 PAP 方式认证对端路由器 Router_2

在 Router_1 上创建本地用户名和密码。

[Router_1]**local-user Router_1**
[Router_1-luser-ROUTER_2]**password simple router**
[Router_1-luser-ROUTER_2]**service-type ppp**

在 Router_1 上配置本地验证的方式为 PAP。

[Router_1]**interface Serial 6/0**
[Router_1-Serial6/0]**ppp authentication-mode pap**

（2）在 Router_2 上配置 PAP 验证时发送的用户名和密码

在 Router_2 上配置被对端验证的密码和用户名，这里注意先配置密码和对端的密码不一致，注意看是否能 Ping 通对端主机。

[Router_2]**interface Serial 6/0**
[Router_2-Serial6/0]**ppp pap local-user Router_1 password simple 111**
　　　　　　　　　　　　//密码和 Router_1 上的本地用户密码不一致。

在 Router_1 上 Ping Router_2，看两台路由器是否连通。

[Router_1]**ping 10.2.1.2**
PING 10.2.1.2: 56 data bytes, press CTRL_C to break
Reply from 10.2.1.2: bytes=56 Sequence=1 ttl=255 time=27 ms
Reply from 10.2.1.2: bytes=56 Sequence=2 ttl=255 time=27 ms
Reply from 10.2.1.2: bytes=56 Sequence=3 ttl=255 time=28 ms
Reply from 10.2.1.2: bytes=56 Sequence=4 ttl=255 time=27 ms
Reply from 10.2.1.2: bytes=56 Sequence=5 ttl=255 time=27 ms

--- 10.2.1.2 ping statistics ---
5 packet(s) transmitted
5 packet(s) received
0.00% packet loss
round-trip min/avg/max = 27/27/28 ms

结果显示两台路由器之间能相互 Ping 通。
把 Router_2 上配置被对端验证的密码改为和 Router_1 上的相同。

[Router_2-Serial6/0]**undo ppp pap local-user**
[Router_2-Serial6/0]**ppp pap local-user Router_1 password simple router**

在 Router_1 上 Ping Router_2，结果是可以 Ping 通的。

[Router_1]**ping 10.2.1.2**
PING 10.2.1.2: 56 data bytes, press CTRL_C to break
Reply from 10.2.1.2: bytes=56 Sequence=1 ttl=255 time=27 ms
Reply from 10.2.1.2: bytes=56 Sequence=2 ttl=255 time=27 ms

```
Reply from 10.2.1.2: bytes=56 Sequence=3 ttl=255 time=28 ms
Reply from 10.2.1.2: bytes=56 Sequence=4 ttl=255 time=27 ms
Reply from 10.2.1.2: bytes=56 Sequence=5 ttl=255 time=27 ms

 --- 10.2.1.2 ping statistics ---
5 packet(s) transmitted
5 packet(s) received
0.00% packet loss
round-trip min/avg/max = 27/27/28 ms
```

3. 实验任务三：PPP CHAP 认证配置

本实验要求两台路由器 PPP 采用 CHAP 验证，其中 Router_1 和 Router_2 既为主验证方又为被验证方，实验开始前首先清空 PAP 的配置。

（1）在 Router_1 上配置本地用户名和密码，验证方式 CHAP，Router_1 作为主验证方

```
[Router_1]local-user Router_1
[Router_1-luser- Router_1]password simple router
[Router_1-luser- Router_1]service-type ppp
[Router_1-luser- Router_1]quit
[Router_1]interface Serial 6/0
[Router_1-Serial6/0]ppp authentication-mode chap
[Router_1-Serial6/0]ppp chap user Router_2
```

（2）Router_2 作为被验证方配置

使用本地用户名及密码进行验证。

```
[Router_2]local-user Router_2
[Router_2-luser- Router_2]password simple router
[Router_2-luser- Router_2]service-type ppp
[Router_2-luser- Router_2]quit
[Router_2]interface Serial 6/0
[Router_2-Serial6/0]ppp chap user Router_1
[Router_2-Serial6/0]shutdown
[Router_2-Serial6/0]undo shutdown
```

（3）检查连通性

在 Router_1 上 Ping Router_2：

```
[Router_1]ping 10.2.1.2
PING 10.2.1.2: 56  data bytes, press CTRL_C to break
Reply from 10.2.1.2: bytes=56 Sequence=1 ttl=255 time=27 ms
Reply from 10.2.1.2: bytes=56 Sequence=2 ttl=255 time=27 ms
Reply from 10.2.1.2: bytes=56 Sequence=3 ttl=255 time=27 ms
Reply from 10.2.1.2: bytes=56 Sequence=4 ttl=255 time=27 ms
Reply from 10.2.1.2: bytes=56 Sequence=5 ttl=255 time=27 ms

 --- 10.2.1.2 ping statistics ---
```

```
5 packet(s) transmitted
5 packet(s) received
0.00% packet loss
round-trip min/avg/max = 27/27/27 ms
```

下面以 Router_2 作为主验证方，Router_1 作为被验证方进行 PPP 验证配置。

（4）在 Router_2 上配置本地用户名和密码，验证方式 CHAP。

```
[Router_2]local-user Router
[Router_2-luser-Router_]password simple router
[Router_2-luser-Router_]service-type ppp
[Router_2-luser-Router_]quit
[Router_2]interface Serial 6/0
[Router_2-Serial6/0]ppp authentication-mode chap
```

（5）Router_1 作为被验证方配置

实验默认 CHAP 密码进行验证。

```
[Router_1]interface Serial 6/0
[Router_1-Serial6/0]ppp chap user Router
[Router_1-Serial6/0]ppp chap password simple 222
```

结果发现不能通过验证，现象同实验任务二 PAP 验证时密码配错一样。
更改 Router_1 接口 S6/0 上 CHAP 验证密码。

```
[Router_1-Serial6/0]ppp chap password simple router
```

（6）测试 Router_1 和 Router_2 的连通性

在 Router_1 上把 S6/0 接口 DOWN 掉再 UP，然后测试连通性。

```
[Router_1]ping 10.2.1.2
PING 10.2.1.2: 56  data bytes, press CTRL_C to break
Reply from 10.2.1.2: bytes=56 Sequence=1 ttl=255 time=27 ms
Reply from 10.2.1.2: bytes=56 Sequence=2 ttl=255 time=27 ms
Reply from 10.2.1.2: bytes=56 Sequence=3 ttl=255 time=27 ms
Reply from 10.2.1.2: bytes=56 Sequence=4 ttl=255 time=27 ms
Reply from 10.2.1.2: bytes=56 Sequence=5 ttl=255 time=28 ms
--- 10.2.1.2 ping statistics ---
5 packet(s) transmitted
5 packet(s) received
0.00% packet loss
round-trip min/avg/max = 27/27/28 ms
```

六、实验思考

① 在做 PPP PAP 认证配置时，两端设备密码不一致为什么还能 Ping 通？怎么解决？
② PAP 认证和 CHAP 认证有何区别？在那些情况下适合用 PAP 单向验证？
③ 在 CHAP 认证配置中为什么要求用户名必须为对端路由器的名称，而且密码必须一致？
④ 在配置验证时可不可以选择同时使用 PAP 和 CHAP？

实验十三　帧中继配置

一、实验目的

① 熟悉帧中继的基本配置；
② 熟悉帧中继网络中 RIP 的配置。

二、实验内容

① 按图 13-1 连接实验设备；
② 设备接口封装帧中继协议、配置好 IP 地址，并查看相应接口信息；
③ 配置路由器模拟帧中继交换机，并检测帧中继交换机是否配置成功；
④ 在帧中继网络上配置 RIP，使用单播和广播两种方法实现，并进行相应的测试。

三、实验设备与组网图

1. 实验设备

三台 H3C 路由器，一条专用 Console 配置线缆，一台计算机，超级终端或 Secure CRT 软件。

2. 组网图

帧中继（frame relay）是一种统计复用的协议，在数据链路层用简化的方法传送和交换数据单元的快速分组交换技术，它能够在单一物理传输线路上提供多条虚电路。

实验组网如图 13-1 所示，实验使用三台 MSR 系列路由（Router_1、Router_2、Router_3）构建一个帧中继网络，其中 Router_2 作为帧中继交换机。Router_1 通过 DLCI 40 标识的 PVC 连接 Router_3，而在 Router_3 一端，此 PVC 通过 DLCI 50 标识。

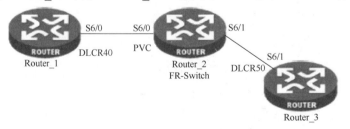

图 13-1　三台路由器用串口相连

四、实验相关知识

1．帧中继协议简介

帧中继协议是一种简化的 X.25 广域网协议。帧中继协议是一种统计复用的协议，它在单一物理传输线路上能够提供多条虚电路。每条虚电路用数据链路连接标识来标识，DLCI 只在本地接口和与之直接相连的对端接口有效，不具有全局有效性，即在帧中继网络中，不同的物理接口上相同的 DLCI 并不表示是同一个虚电路。

帧中继是一种局域网互联的 WAN 协议，它工作在 OSI 参考模型的物理层和数据链路层。它为跨越多个交换机和路由器的用户设备间的信息传输提供了快速和有效的方法。

帧中继是一种数据包交换技术，与 X.25 类似。它可以使终端站动态共享网络介质和可用带宽。帧中继采用两种数据包技术，可变长数据包和统计多元技术。它不能确保数据完整性，所以当出现网络拥塞现象时，就会丢弃数据包。但在实际应用中，它仍然具有可靠的数据传输性能。

帧中继网络既可以是公用网络或者某一企业的私有网络，也可以是数据设备之间直接连接构成的网络。

2．帧中继的特点

帧中继技术的特点如下。

① 帧中继技术主要用于传递数据信息，它将数据信息以满足帧中继协议的帧的形式有效地进行传送。

② 帧中继传送数据信息所使用的传输链路是逻辑连接，而不是物理连接。在一个物理连接上可以复用多个逻辑连接，使用这种方式可实现带宽复用及动态分配带宽。

③ 帧中继协议简化了 X.25 的第三层功能，使网络功能的处理大大简化，提高了网络对信息处理的效率。只采用物理层和链路层的两级结构，在链路层中仅保留其核心的子集部分。

④ 在链路层完成统计复用、帧透明传输和错误检测，但不提供发现错误后的重传操作，省去了帧编号、流量控制、应答和监视等机制，大大节省了交换机的开销，提高了网络吞吐量、降低了通信时延。一般 FR 用户的接入速率在 64kbits/s～2Mbits/s 之间，近期 FR 的速率已提高到 8～10Mbits/s，今后将达到 45Mbits/s。

⑤ 交换单元——帧的信息长度远比分组长度要长，预约的最大帧长度至少要达到 1600B/帧，适合于封装局域网（LAN）的数据单元。

⑥ 提供一套合理的带宽管理和防止阻塞的机制，用户有效地利用预先约定的带宽，即承诺的信息速率（CIR），并且还允许用户的突发数据占用未预定的带宽，以提高整个网络资源的利用率。

⑦ 与分组交换一样，FR 采用面向连接的交换技术，可以提供交换虚电路 SVC 业务和永久虚电路 PVC 业务，但目前已应用的 FR 网络中只采用 PVC 业务。

3．帧中继地址影射

帧中继地址映射是把对端设备的协议地址与对端设备的帧中继地址（本地的 DLCI）

关联起来，使高层协议能通过对端设备的协议地址寻址到对端设备。

帧中继主要用来承载 IP 协议，在发送 IP 报文时，根据路由表只能知道报文的下一跳地址，发送前必须由该地址确定它对应的 DLCI。这个过程可以通过查找帧中继地址映射表来完成，因为地址映射表中存放的是下一跳 IP 地址和下一跳对应的 DLCI 的映射关系。

地址映射表可以由手工配置，也可以由逆向地址解析协议 Inverse ARP 动态维护。

（1）帧中继 DLCI 的手工分配

从帧中继网络服务商处得到分配的 DLCIs。每个 DLCI 只有本地意义。映射对端的网络地址到 DLCIs。

（2）Inverse ARP

Inverse ARP 可以自动发现目的路由器的网络地址，从而简化了帧中继的配置。

五、实验过程

1. 实验任务一：配置帧中继网络

（1）搭建实验环境，初始化实验设备

按图 13-1 所示搭建实验环境，清除配置文件，以出厂配置重启网络设备。

```
<H3C>reset saved-configuration
<H3C>reboot
……
<H3C>
```

（2）按表 13-1 所示更改设备名称并配置 PC、路由器端口 IP 地址

实验要求在 Router_1 上配置子接口，帧中继配置完成后，Router_1 能 Ping 通 Router_3。

表 13-1　IP 地址规划表

设备	接口	IP 地址/掩码
Router_1	S6/0.2	10.1.1.1/30
Router_3	S6/1	10.1.1.2/30

（3）将 Router_1 接口封装帧中继，并配置帧中继相关参数

```
[Router_1]interface Serial 6/0
[Router_1-Serial6/0]link-protocol fr          //封装帧中继类型。
[Router_1-Serial6/0]fr interface-type dte     //默认情况下接口类型为 DTE。
[Router_1-Serial6/0]fr lmi type q933a         //默认情况下 LMI 类型为 q933a。
[Router_1-Serial6/0]quit
[Router_1]interface s6/0.2 p2mp
[Router_1-Serial6/0.2]fr dlci 40   //子接口下要配置 DLCI 号，而物理接口下可不配。
[Router_1-Serial6/0.2]ip address 10.1.1.1 30
```

```
[Router_1-Serial6/0.2]fr map ip 10.1.1.2 40
```

(4) 验证 Router_1 接口配置的正确性

在 Router_1 上通过 display　fr map-infp 查看帧中继配置。

```
[Router_1]display fr map-info
PVC map Statistics for interface Serial6/0（DTE）          //接口类型为 DTE。
DLCI = 40, IP 10.1.1.2, Serial6/0.2
                    //地址映射：DLCI 号 40，IP 地址 10.1.1.2，接口为 s1/0.2。
create time = 2011/04/02 08:42:04, status = INACTIVE
encapsulation = ietf, vlink = 0
[ROUTER_1]display fr lmi-info
Frame relay LMI statistics for interface Serial6/0  DTE, Q933
                                       //LMI 类型为 q922a。
T391DTE = 10（hold timer 10）
N391DTE = 6, N392DTE = 3, N393DTE = 4
out status enquiry = 168, in status = 91
status timeout = 74, discarded messages = 2
```

(5) 将 Router_3 接口封装帧中继，并配置帧中继相关参数

```
[Router_3]interface Serial 6/1
[Router_3-Serial6/1]link-protocol fr
[Router_3-Serial6/1]fr interface-type dte
[Router_3-Serial6/1]fr lmi type q933a
[Router_3-Serial6/1]ip address 10.1.1.2 30
[Router_3-Serial6/1]fr map ip 10.1.1.2 50
```

(6) 验证 Router_3 接口配置的正确性

在 Router_3 上通过 display fr map-infp 查看帧中继配置。

```
[Router_3]display fr map-info
Map Statistics for interface Serial6/1（DTE）
DLCI = 50, IP 10.1.1.2, Serial6/1
create time = 2016/07/03 14:58:50, status = INACTIVE
encapsulation = ietf, vlink = 0
[Router_3]display fr lmi-info
Frame relay LMI statistics for interface Serial6/1（DTE, Q933）
T391DTE = 10（hold timer 10）
N391DTE = 6, N392DTE = 3, N393DTE = 4
out status enquiry = 14, in status = 0
status timeout = 13, discarded messages = 0
```

(7) 配置 Router_2 模拟帧中继交换机

```
[Router_2]fr switching
[Router_2]interface Serial 6/0
[Router_2-Serial6/0]link-protocol fr
```

```
[Router_2-Serial6/0]fr interface-type dce
[Router_2-Serial6/0]fr lmi type q933a
[Router_2-Serial6/0]fr dlci 40              //给对端分配的DLCI号。
[Router_2-fr-dlci-Serial6/0-40]quit
[Router_2-Serial6/0]quit
[Router_2]interface Serial 6/1
[Router_2-Serial6/1]link-protocol fr
[Router_2-Serial6/1]fr interface-type dce
[Router_2-Serial6/1]fr lmi type q933a
[Router_2-Serial6/1]fr dlci 50
[Router_2-fr-dlci-Serial6/1-50]quit
[Router_2-Serial6/1]quit
[Router_2]fr switch a-c interface Serial 6/0 dlci 40 interface Serial 6/1 dlci 50                //Router_1与Router_3间的PVC。
```

检测帧中继交换机是否配置成功：

```
[Router_2]display fr switch-table all
Total PVC switch records:1
PVC-Name            Status   Interface（DLCI）<----->   Interface（Dlci）
a-c                 Active   Serial6/0（40）             Serial6/1（50）
```

（8）验证 ROUTER_1 和 ROUTER_3 的连通性

```
[Router_1]ping 10.1.1.2
PING 10.1.1.2: 56  data bytes, press CTRL_C to break
Reply from 10.1.1.2: bytes=56 Sequence=1 ttl=255 time=53 ms
Reply from 10.1.1.2: bytes=56 Sequence=2 ttl=255 time=53 ms
Reply from 10.1.1.2: bytes=56 Sequence=3 ttl=255 time=53 ms
Reply from 10.1.1.2: bytes=56 Sequence=4 ttl=255 time=52 ms
Reply from 10.1.1.2: bytes=56 Sequence=5 ttl=255 time=53 ms

--- 10.1.1.2 ping statistics ---
5 packet(s) transmitted
5 packet(s) received
0.00% packet loss
round-trip min/avg/max = 52/52/53 ms
```

2. 实验任务二：在帧中继网络上配置 RIP

实验任务一完成后，确认 Router_1 和 Router_3 能互通后，在 Router_1 和 Router_3 上配置 RIP。为了便于测试，在 Router_1 上配置 LoopBack 0 地址为 20.1.1.1/32，在 Router_3 上配置 LoopBack 0 地址为 30.1.1.1/32。

（1）在 Router_1 和 Router_3 上添加配置 RIP 协议

Router_1 上的 RIP 配置：

```
[Router_1]interface LoopBack 0
[Router_1-LoopBack0]ip address 20.1.1.1 32
```

```
[Router_1]rip
[Router_1-rip]version 2
[Router_1-rip]network 20.1.1.1
[Router_1-rip]network 10.0.0.0
```

Router_3 上的 RIP 配置：

```
[Router_3]interface LoopBack 0
[Router_3-LoopBack0]ip address 30.1.1.1 32
[Router_3]rip
[Router_3-rip]version 2
[Router_3-rip]network 30.1.1.1
[Router_3-rip]network 10.0.0.0
```

（2）查看路由表信息

在 Router_3 上查看路由表：

```
[Router_3]display ip routing-table
Routing Tables: Public
Destinations : 5        Routes : 5
Routing Table: public net
Destination/Mask   Protocol  Pre  Cost     Nexthop      Interface
10.1.1.0/30        DIRECT    0    0        10.1.1.2     Serial6/1
10.1.1.1/32        DIRECT    0    0        10.1.1.1     Serial6/1
10.1.1.2/32        DIRECT    0    0        127.0.0.1    InLoopBack0
30.1.1.1/32        DIRECT    0    0        127.0.0.1    InLoopBack0
127.0.0.0/8        DIRECT    0    0        127.0.0.1    InLoopBack0
127.0.0.0/32       DIRECT    0    0        127.0.0.1    InLoopBack0
```

结果显示在 Router_3 上没有收到 RIP 路由信息。这里再采用 debug rip packet：

```
< Router_3>debugging rip packet
  Rip packet debugging is on
< Router_3>
*0.5834902 ROUTER_3 RM/7/RTDBG:
RIP: send from 10.1.1.2(Serial6/1) to 224.0.0.9
  Packet:vers 2, cmd Response, length 24
  dest 30.1.1.1   mask 255.255.255.255, router 0.0.0.0,    metric 1, tag 0
```

结果显示在 Router_3 上只有发出的 RIP 报文而没有收到的 RIP 报文。

下面针对这种情况分别用两种方法进行配置。

方法一：在 Router_1 上修改 RIP 配置。

```
[Router_1]rip
[Router_1-rip]peer 10.1.1.2         //以单播方式发送RIP更新报文给10.1.1.2。
```

在 Router_3 上修改 RIP 配置。

```
[Router_3]rip
[Router_3-rip]peer 10.1.1.1
```

查看路由表信息：

```
[Router_3]display ip routing-table
Routing Table: public net
Destination/Mask    Protocol    Pre    Cost    Nexthop        Interface
10.1.1.0/30         DIRECT      0      0       10.1.1.2       Serial6/1
10.1.1.1/32         DIRECT      0      0       10.1.1.1       Serial6/1
10.1.1.2/32         DIRECT      0      0       127.0.0.1      InLoopBack0
20.1.1.1/32         RIP         100    1       10.1.1.1       Serial6/1
30.1.1.1/32         DIRECT      0      0       127.0.0.1      InLoopBack0
127.0.0.0/8         DIRECT      0      0       127.0.0.1      InLoopBack0
127.0.0.1/32        DIRECT      0      0       127.0.0.1      InLoopBack0
```

在 Router_3 上测试连通性：

```
[Router_3]ping -a 30.1.1.1 20.1.1.1
PING 20.1.1.1: 56  data bytes, press CTRL_C to break
 Reply from 20.1.1.1: bytes=56 Sequence=1 ttl=255 time=53 ms
 Reply from 20.1.1.1: bytes=56 Sequence=2 ttl=255 time=53 ms
 Reply from 20.1.1.1: bytes=56 Sequence=3 ttl=255 time=53 ms
 Reply from 20.1.1.1: bytes=56 Sequence=4 ttl=255 time=53 ms
 Reply from 20.1.1.1: bytes=56 Sequence=5 ttl=255 time=54 ms
--- 20.1.1.1 ping statistics ---
5 packet(s) transmitted
5 packet(s) received
0.00% packet loss
round-trip min/avg/max = 53/53/54 ms
```

查看 debug rip packet：

```
< Router_3>debugging rip packet
  Rip packet debugging is on
< Router_3>
*0.6914832 ROUTER_3 RM/7/RTDBG:
  RIP: send from 10.1.1.2(Serial6/1) to 224.0.0.9
Packet:vers 2, cmd Response, length 24
dest 30.1.1.1    mask 255.255.255.255, router 0.0.0.0,    metric 1, tag 0
*0.6914832 ROUTER_3 RM/7/RTDBG:
RIP: send from 10.1.1.2(Serial6/1) to 10.1.1.1
Packet:vers 2, cmd Response, length 24
dest 30.1.1.1    mask 255.255.255.255, router 0.0.0.0,    metric 1, tag 0
*0.6917970 ROUTER_3 RM/7/RTDBG:
RIP: Receive Response from 10.1.1.1+520 via Serial6/1(10.1.1.2)
Packet:vers 2, cmd Response, length 24
dest 20.1.1.1    mask 255.255.255.255, router 0.0.0.0 , metric 1, tag 0
```

可以发现：Router_3 发送报文使用单播目的地址 10.1.1.2，同时也接收来自 10.1.1.1 的

RIP 报文。说明在帧中继网络中可以通过配置 RIP peer 来实现路由器学习 RIP 路由信息。

方法二：修改 Router_1 上静态映射。

[Router_1]**interface Serial 6/0.2**
[Router_1-Serial6/0.2]**undo fr map ip 10.1.1.2 40**
[Router_1-Serial6/0.2]**fr map ip 10.1.1.2 40 broadcast**
　　　　　　　　　//可以用广播传送报文修改 Router_3 上静态映射。
[Router_3]**interface Serial 6/1**
[Router_3-Serial6/1]**undo fr map ip 10.1.1.1 50**
[Router_3-Serial6/1]**fr map ip 10.1.1.1 50 broadcast**

查看路由表信息：

[Router_1]**display ip routing-table**
Routing Table: public net

Destination/Mask	Protocol	Pre	Cost	NextHop	Interface
10.1.1.0/30	DIRECT	0	0	10.1.1.1	Serial6/0.2
10.1.1.1/32	DIRECT	0	0	127.0.0.1	InLoopBack0
10.1.1.2/32	DIRECT	0	0	10.1.1.2	Serial6/0.2
20.1.1.1/32	DIRECT	0	0	127.0.0.1	InLoopBack0
30.1.1.1/32	**RIP**	**100**	**1**	**10.1.1.2**	**Serial6/0.2**
127.0.0.0/8	DIRECT	0	0	127.0.0.1	InLoopBack0
127.0.0.1/32	DIRECT	0	0	127.0.0.1	InLoopBack0

[Router_3]**display ip routing-table**
Routing Table: public net

Destination/Mask	Protocol	Pre	Cost	NextHop	Interface
10.1.1.0/30	DIRECT	0	0	10.1.1.2	Serial6/1
10.1.1.1/32	DIRECT	0	0	10.1.1.1	Serial6/1
10.1.1.2/32	DIRECT	0	0	127.0.0.1	InLoopBack0
20.1.1.1/32	**RIP**	**100**	**1**	**10.1.1.1**	**Serial6/1**
30.1.1.1/32	DIRECT	0	0	127.0.0.1	InLoopBack0
127.0.0.0/8	DIRECT	0	0	127.0.0.1	InLoopBack0
127.0.0.1/32	DIRECT	0	0	127.0.0.1	InLoopBack0

在 Router_1 上测试连通性：

[Router_1]**ping -a 20.1.1.1 30.1.1.1**
　PING 30.1.1.1: 56 data bytes, press CTRL_C to break
　　Reply from 30.1.1.1: bytes=56 Sequence=1 ttl=255 time=53 ms
　　Reply from 30.1.1.1: bytes=56 Sequence=2 ttl=255 time=52 ms
　　Reply from 30.1.1.1: bytes=56 Sequence=3 ttl=255 time=53 ms
　　Reply from 30.1.1.1: bytes=56 Sequence=4 ttl=255 time=53 ms
　　Reply from 30.1.1.1: bytes=56 Sequence=5 ttl=255 time=53 ms
　--- 30.1.1.1 ping statistics ---
　　5 packet(s) transmitted
　　5 packet(s) received
　　0.00% packet loss
　　round-trip min/avg/max = 52/52/53 ms

查看 debug rip packet：

```
< Router_1>debugging rip packet
Rip packet debugging is on
< Router_1>
*0.7637249 ROUTER_1 RM/7/RTDBG:
RIP: Receive Response from 10.1.1.2+520 via Serial6/0.2 (224.0.0.9)
 Packet:vers 2, cmd Response, length 24
 dest 30.1.1.1    mask 255.255.255.255, router 0.0.0.0, metric 1, tag 0
*0.7640242 ROUTER_1 RM/7/RTDBG:
 RIP: send from 10.1.1.1(Serial6/0.2) to 224.0.0.9
 Packet:vers 2, cmd Response, length 24
 dest 20.1.1.1   mask 255.255.255.255, router 0.0.0.30, metric 1, tag 0
```

可以发现 Router_1 发送报文使用广播目的地址 10.1.1.1，同时也接收来自 10.1.1.2 的 RIP 报文。说明在帧中继网络中可以通过配置静态映射发送广播报文来实现路由器学习 RIP 路由信息。

六、实验思考

① 在帧中继配置中为什么要用子接口？它有哪些分类？

② 帧中继 LMI 类型有哪些？分别怎么用？

③ 在开始配置 RIP 时结果显示在 Router_3 上，只有发出的 RIP 报文而没有收到 RIP 报文。说明这时 RIP 协议没有工作起来。思考为什么？

实验十四　ACL 包过滤

一、实验目的

① 了解访问控制列表的简单工作原理；
② 掌握基本 ACL 的配置；
③ 掌握高级 ACL 的配置。

二、实验内容

① 按图 14-1 规划地址，连接实验设备，并进行连通性测试；
② 配置基本 ACL，并进行验证；
③ 配置高级 ACL，并进行验证。

三、实验设备与组网图

1．实验设备

两台 H3C 路由器，一条专用 Console 配置线缆，一台计算机，超级终端或 Secure CRT 软件。

2．组网图

用网线将两台路由器以太网口相连，配置 ACL，以实现 Router_1 上不能 Ping 通 Router_2，如图 14-1 所示。

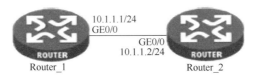

图 14-1　路由器之间用以太网口相连

四、实验相关知识

1．ACL 介绍

访问控制列表（access control list，ACL）是用来实现流识别功能的。网络设备为了过滤报文，需要配置一系列匹配条件对报文进行分类，这些条件可以是报文的源地址、目的地址、端口号等。当设备的端口接收到报文后，即根据当前端口上应用的 ACL 规则对报文的字段进行分析，在识别出特定的报文之后，根据预先设定的策略允许或禁止该报文通过。

由 ACL 定义的报文匹配规则，可以被其他需要对流量进行区分的场合引用，如包过滤、QoS 中流分类规则的定义等。ACL 根据 ACL 序号来区分不同的 ACL，可以分为四种类型，如表 14-1 所示。

表 14-1 ACL 类型表

ACL 类型	ACL 序号范围	区分报文的依据
基本 ACL	2000～2999	只根据报文的源 IP 地址信息制定匹配规则
高级 ACL	3000～3999	根据报文的源 IP 地址信息、目的 IP 地址信息、IP 承载的协议类型、协议的特性等三、四层信息制定匹配规则
二层 ACL	4000～4999	根据报文的源 MAC 地址、目的 MAC 地址、802.1p 优先级、二层协议类型等二层信息制定匹配规则
用户自定义 ACL	5000～5999	可以以报文的报文头、IP 头等为基准，指定从第几个字节开始与掩码进行"与"操作，将从报文提取出来的字符串和用户定义的字符串进行比较，找到匹配的报文

用户在创建 ACL 时，可以为 ACL 指定一个名称。每个 ACL 最多只能有一个名称。命名的 ACL 使用户可以通过名称唯一地确定一个 ACL，并对其进行相应的操作。

在创建 ACL 时，用户可以选择是否配置名称。ACL 创建后，不允许用户修改或者删除，也不允许为未命名的 ACL 添加名称。

2．ACL 的使用

防火墙配置常见步骤：
① 启用防火墙；
② 定义访问控制列表；
③ 将访问控制列表应用到接口上。

3．ACL 的作用

访问控制列表可以用于防火墙：
① 访问控制列表可以用于 Qos（quality of service），对数据流量进行控制；
② 在 DCC 中，访问控制列表还可用来规定触发拨号的条件；
③ 访问控制列表还可以用于地址转换；
④ 在配置路由策略时，可以通过访问控制列表进行路由信息的过滤。

4．ACL 匹配顺序

一个 ACL 中可以包含多个规则，而每个规则都指定不同的报文匹配选项，这些规则可能存在重复或矛盾的地方，在将一个报文和 ACL 的规则进行匹配的时候，到底采用哪些规则呢？这就需要确定规则的匹配顺序。

ACL 支持两种匹配顺序。

配置顺序：按照用户配置规则的先后顺序进行规则匹配。

自动排序：按照"深度优先"的顺序进行规则匹配。

（1）基本 ACL 的"深度优先"顺序判断原则

① 先看规则中是否带 VPN 实例，带 VPN 实例的规则优先；
② 再比较源 IP 地址范围，源 IP 地址范围小，反掩码中"0"位的数量多的规则优先；
③ 如果源 IP 地址范围相同，则先配置的规则优先。

（2）高级 ACL 的"深度优先"顺序判断原则

① 先看规则中是否带 VPN 实例，带 VPN 实例的规则优先；
② 再比较协议范围，指定了 IP 协议承载的协议类型的规则优先；
③ 如果协议范围相同，则比较源 IP 地址范围，源 IP 地址范围小，反掩码中"0"位的数量多的规则优先；
④ 如果协议范围、源 IP 地址范围相同，则比较目的 IP 地址范围，目的 IP 地址范围小，反掩码中"0"位的数量多的规则优先；
⑤ 如果协议范围、源 IP 地址范围、目的 IP 地址范围相同，则比较四层端口号（TCP/UDP 端口号）范围，四层端口号范围小的规则优先；
⑥ 如果上述范围都相同，则先配置的规则优先。

（3）二层 ACL 的"深度优先"顺序判断原则

① 先比较源 MAC 地址范围，源 MAC 地址范围小，掩码中"1"位的数量多的规则优先；
② 如果源 MAC 地址范围相同，则比较目的 MAC 地址范围，目的 MAC 地址范围小，掩码中"1"位的数量多的规则优先；
③ 如果源 MAC 地址范围、目的 MAC 地址范围相同，则先配置的规则优先。

在报文匹配规则时，会按照匹配顺序去匹配定义的规则，一旦有一条规则被匹配，报文就不再继续匹配其他规则了，设备将对该报文执行第一次匹配的规则指定的动作。

五、实验过程

1. 实验任务一：用基本 ACL 实现

（1）搭建实验环境，初始化实验设备

按图 14-1 所示搭建实验环境，清除配置文件，以出厂配置重启网络设备。

```
<H3C>reset saved-configuration
<H3C>reboot
……
<H3C>
```

（2）按图 14-1 所示更改设备名并配置 PC、路由器端口 IP 地址，测试连通性

在 Router_1 上 Ping Router_2，显示如下：

```
[Router_1]ping 10.1.1.2
 PING 10.1.1.2: 56  data bytes, press CTRL_C to break
    Reply from 10.1.1.2: bytes=56 Sequence=1 ttl=255 time=1 ms
    Reply from 10.1.1.2: bytes=56 Sequence=2 ttl=255 time=1 ms
    Reply from 10.1.1.2: bytes=56 Sequence=3 ttl=255 time=1 ms
    Reply from 10.1.1.2: bytes=56 Sequence=4 ttl=255 time=1 ms
    Reply from 10.1.1.2: bytes=56 Sequence=5 ttl=255 time=1 ms

 --- 10.1.1.2 ping statistics ---
  5 packet(s) transmitted
  5 packet(s) received
  0.00% packet loss
round-trip min/avg/max = 1/1/1 ms
```

在 Router_2 上开启 telnet server。

```
<Router_1>telnet 10.1.1.2 23
Trying 10.1.1.2 ...
Press CTRL+K to abort
Connected to 10.1.1.2 ...
******************************************************************
* Copyright (c) 2004-2009 Hangzhou H3C Tech. Co., Ltd. All rights reserved. *
* Without the owner's prior written consent,                      *
* no decompiling or reverse-engineering shall be allowed.         *
******************************************************************

Welcome!
Login authentication

Username:h3c
Password:
<Router_2>
```

（3）创建基本 ACL

```
[Router_2]acl number 2000
[Router_2-acl-basic-2000]rule 0 deny source 10.1.1.1 0.0.0.0
```

（4）应用 ACL

基本 ACL 应部署在靠近目的的地方。因此，部署在 Router_2 的 GE0/0。

```
[Router_2]firewall enable
[Router_2]firewall default permit
[Router_2]interface GigabitEthernet 0/0
[Router_2-GigabitEthernet 0/0]firewall packet-filter ?
  INTEGER<2000-2999>  Apply basic acl
  INTEGER<3000-3999>  Apply advanced acl
```

```
  INTEGER<4000-4999>  Apply ethernet frame header acl
  ipv6                ACL IPv6
  name                Specify a named acl

[Router_2-GigabitEthernet 0/0]firewall packet-filter 2000 ?
  inbound   Apply the acl to filter in-bound packets
  outbound  Apply the acl to filter out-bound packets

[Router_2-GigabitEthernet 0/0]firewall packet-filter 2000 inbound ?
  <cr>
[Router_2-GigabitEthernet 0/0]firewall packet-filter 2000 inbound
```

（5）验证

用在 Router_1 上 Ping 路由器 Router_2：

```
[Router_1]ping 10.1.1.2
  PING 10.1.1.2: 56  data bytes, press CTRL_C to break
    Request time out
    Request time out
    Request time out
    Request time out
    Request time out

  --- 10.1.1.2 ping statistics ---
    5 packet(s) transmitted
    0 packet(s) received
    100.00% packet loss
```

远程登录 Router_2：

```
<Router_1>telnet 10.1.1.2 23
Trying 10.1.1.2 ...
Press CTRL+K to abort
Can't connect to the remote host!
```

2. 实验任务二：用高级 ACL 实现

（1）搭建实验环境，初始化实验设备

按图 14-1 所示搭建实验环境，清除配置文件，以出厂配置重启网络设备。

```
<H3C>reset saved-configuration
<H3C>reboot
……
<H3C>
```

（2）更改设备名称并配置地址

按图 14-1 所示更改设备名称并配置 PC、路由器端口 IP 地址，测试连通性在 Router_1 上 Ping Router_2，显示如下：

```
[Router_1]ping 10.1.1.2
  PING 10.1.1.2: 56  data bytes, press CTRL_C to break
    Reply from 10.1.1.2: bytes=56 Sequence=1 ttl=255 time=1 ms
    Reply from 10.1.1.2: bytes=56 Sequence=2 ttl=255 time=1 ms
    Reply from 10.1.1.2: bytes=56 Sequence=3 ttl=255 time=1 ms
    Reply from 10.1.1.2: bytes=56 Sequence=4 ttl=255 time=1 ms
    Reply from 10.1.1.2: bytes=56 Sequence=5 ttl=255 time=1 ms

  --- 10.1.1.2 ping statistics ---
    5 packet(s) transmitted
    5 packet(s) received
    0.00% packet loss
    round-trip min/avg/max = 1/1/1 ms
```

（3）创建高级 ACL

```
[Router_1]acl number 3000
[Router_1-acl-adv-3000]rule 0 deny icmp source 10.1.1.1 0.0.0.0 destination 10.1.1.2 0.0.0.0
```

（4）部署 ACL

高级 ACL 应部署在靠近源的地方。因此，部署在 Router_1 的 GE0/0。

```
[Router_1]firewall enable
[Router_1]firewall default permit
[Router_1]interface GigabitEthernet 0/0
[Router_1-GigabitEthernet0/0]firewall packet-filter 3000 outbound
```

（5）验证

```
[Router_1]ping 10.1.1.2
  PING 10.1.1.2: 56  data bytes, press CTRL_C to break
    Request time out
    Request time out
    Request time out
    Request time out
    Request time out

  --- 10.1.1.2 ping statistics ---
    5 packet(s) transmitted
    0 packet(s) received
    100.00% packet loss
```

显示配置的 ACL：

```
[Router_1]display acl 3000
Advanced ACL  3000, named -none-, 1 rule,
ACL's step is 5
```

```
  rule 0 deny icmp source 10.1.1.1 0 destination 10.1.1.2 0 ( 5 times
matched )
```

查看防火墙的统计信息：

```
[Router_1]display firewall-statistics all
  Firewall is enable, default filtering method is 'permit'.

  Interface: GigabitEthernet0/0
  Out-bound Policy: acl 3000
  Fragments matched normally
  From 2016-07-03 16:37:20 to 2016-07-03 16:37:53
     0 packets, 0 bytes, 0% permitted,
     5 packets, 420 bytes, 100% denied,
     0 packets, 0 bytes, 0% permitted default,
     0 packets, 0 bytes, 0% denied default,
  Totally 0 packets, 0 bytes, 0% permitted,
  Totally 5 packets, 420 bytes, 100% denied.
```

远程登录测试：

```
<Router_1>telnet 10.1.1.2 23
Trying 10.1.1.2 ...
Press CTRL+K to abort
Connected to 10.1.1.2 ...
**********************************************************************
* Copyright ( c ) 2004-2009 Hangzhou H3C Tech. Co., Ltd. All rights
reserved.  *
* Without the owner's prior written consent,                          *
* no decompiling or reverse-engineering shall be allowed.             *
**********************************************************************
Welcome!
Login authentication
Username:
Password:
```

六、实验思考

① 在基本 ACL 配置中应用基本 ACL 后为什么 TELNET 也不通？
② 在 ACL 配置中"inbound"和"outbound"的应用有什么区别？分别怎么用？
③ 在配置高级 ACL 时为什么把高级访问控制列表放在靠近过滤源的位置上？

实验十五 NAT 配置

一、实验目的

① 掌握 Basic NAT 的配置方法；
② 掌握 NAPT 的配置方法；
③ 掌握 Easy IP 的配置方法；
④ 掌握 NAT Server 的配置方法。

二、实验内容

① 按图 15-1 规划地址，连接实验设备，并进行连通性测试；
② 配置 Basic NAT，进行测试验证，并显示和调试 NAT 信息；
③ 配置 NAPT，进行测试验证，并检查 NAT 表项；
④ 配置 Easy IP，进行测试验证，并检查 NAT 表项；
⑤ 配置 NAT Server，进行测试验证，并检查 NAT 表项。

三、实验设备与组网图

1．实验设备

三台 H3C 路由器，一条专用 Console 配置线缆，两台计算机，超级终端或 Secure CRT 软件。

2．组网图

实验组网如图 15-1 所示，实验使用两台主机（PCA，PCB）、三台 MSR 路由器。PCA、PCB 位于私网，网关为 Router_1。Router_2 为 NAT 设备，有一个私网接口 S6/0 和一个公网接口 GE0/0，公网接口与公网路由器 Router_3 互连，地址池范围是 200.1.1.11 至 200.1.1.21。公网 Router_3 上 LoopBack 0 是一个公网接口地址，作为公网服务器测试。

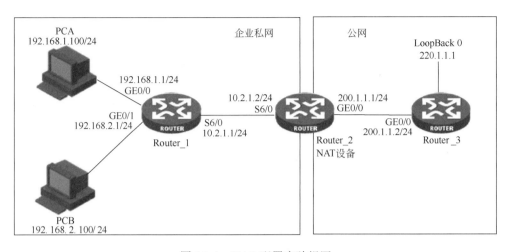

图 15-1　NAT 配置实验组网

四、实验相关知识

1．NAT 概述

网络地址转换（network address translation，NAT）是将 IP 数据报文报头中的 IP 地址转换为另一个 IP 地址的过程。在实际应用中，NAT 主要用于实现私有网络访问公共网络的功能。这种通过使用少量的公有 IP 地址代表较多的私有 IP 地址的方式，将有助于减缓可用 IP 地址空间的枯竭。

IP 地址耗尽促成了 CIDR 的开发，但 CIDR 开发的主要目的是有效地使用现有的 Internet 地址。而同时根据 RFC 1631 开发的 NAT 却可以在多重 Internet 子网中使用相同的 IP，用来减少注册 IP 地址的使用。

NAT 技术使得一个私有网络可以通过 Internet 注册 IP 连接到外部世界，位于内部网络和外部网络中的 NAT 路由器在发送数据包之前，负责把内部 IP 翻译成外部合法地址。内部网络的主机不可能同时与外部网络通信，所以只有一部分内部地址需要翻译。

NAT 使用的几种情况：
① 连接到 Internet，但却没有足够的合法地址分配给内部主机；
② 更改到一个需要重新分配地址的 ISP；
③ 有相同的 IP 地址的两个 Internet 合并；
④ 想支持主机负载均衡。

地址转换的优点在于，在为内部主机提供了"隐私"保护的前提下，实现了内部网络的主机通过该功能访问外部网络的资源。但它也有以下缺点。

① 由于需要对数据报文进行 IP 地址的转换，涉及 IP 地址的数据报的报头不能被加密。在应用协议中，如果报文中有地址或端口需要转换，则报文不能被加密。例如，不能使用加密的 FTP 连接，否则 FTP 的 port 命令不能被正确转换。

② 网络调试变得更加困难。比如，某一台内部网络的主机试图攻击其他网络，则很难指出究竟哪一台机器是恶意的，因为主机的 IP 地址被屏蔽了。

③ 在链路的带宽低于 1.5Gbits/s 时，地址转换对网络性能影响很小，此时，网络传输的瓶颈在传输线路上；当速率高于 1.5Gbits/s 时，地址转换将对网络性能产生一些影响。

2．NAT 的工作原理

NAT 技术能帮助解决令人头痛的 IP 地址紧缺的问题，而且能使得内、外网络隔离，提供一定的网络安全保障。它解决问题的办法是：在内部网络中使用内部地址，通过 NAT 把内部地址翻译成合法的 IP 地址在 Internet 上使用，其具体的做法是把 IP 包内的地址域用合法的 IP 地址替换。NAT 功能通常被集成到路由器、防火墙、ISDN 路由器或者单独的 NAT 设备中。NAT 设备维护一个状态表，用来把非法的 IP 地址映射到合法的 IP 地址上去。每个包在 NAT 设备中都被翻译成正确的 IP 地址，发往下一级，这意味着给处理器带来了一定的负担。但对于一般的网络来说，这种负担是微不足道的。

3．NAT 技术的类型

NAT 有三种类型：静态 NAT、动态地址 NAT、网络地址端口转换 NAPT。其中静态 NAT 设置起来最为简单和最容易实现，内部网络中的每个主机都被永久映射成外部网络中的某个合法的地址。而动态地址 NAT 则是在外部网络中定义了一系列合法地址，采用动态分配的方法映射到内部网络。NAPT 则是把内部地址映射到外部网络的一个 IP 地址的不同端口上。根据不同的需要，三种 NAT 方案各有利弊。

动态地址 NAT 只是转换 IP 地址，它为每一个内部的 IP 地址分配一个临时的外部 IP 地址，主要应用于拨号，对于频繁的远程连接也可以采用动态 NAT。当远程用户连接上之后，动态地址 NAT 就会分配一个 IP 地址，当用户断开时，这个 IP 地址就会被释放而留待以后使用。

网络地址端口转换 NAPT（network address port translation）是人们比较熟悉的一种转换方式。NAPT 普遍应用于接入设备中，它可以将中小型的网络隐藏在一个合法的 IP 地址后面。NAPT 与动态地址 NAT 不同，它将内部连接映射到外部网络中的一个单独的 IP 地址上，同时在该地址上加上一个由 NAT 设备选定的 TCP 端口号。

在 Internet 中使用 NAPT 时，所有不同的 TCP 和 UDP 信息流看起来似乎来源于同一个 IP 地址。这个优点在小型办公室内非常实用，通过从 ISP 处申请的一个 IP 地址，将多个连接通过 NAPT 接入 Internet。实际上，许多 SOHO 远程访问设备支持基于 PPP 的动态 IP 地址。这样，ISP 甚至不需要支持 NAPT，就可以做到多个内部 IP 地址共用一个外部 IP 地址上 Internet，虽然这样会导致信道的一定拥塞，但考虑到节省的 ISP 上网费用和易治理的特点，用 NAPT 还是很值得的。

4．地址转换及其控制

当内部网络访问外部网络时，NAT 将选择一个合适的公有地址替换内部网络数据包的源地址，如果 NAT 服务器出口上只定义一个 IP 地址。这样当所有的内部网络主机访问外部网络时，只拥有这一个公有 IP 地址。在这种情况下，某一时刻只允许最多一台内部主机访问外部网络，这种情况称为"一对一网络地址转换"。在内部网络的多台主机并发地请求访问外部网络时，"一对一网络地址转换"仅能够实现其中一台主机的访问请求。

NAT 也可实现对并发请求的响应。允许 NAT 服务器拥有多个公有 IP 地址,当第一台内部主机访问外部网络时,NAT 进程选择一个公有地址,并在网络地址转换表中添加记录;当另一台内部主机访问网络时,NAT 进程选择另一个公有地址。依此类推,从而满足了多台内部主机并发访问外部网络的请求,这称为"多对多网络地址转换"。

这两种地址转换方式的特点如下。

(1)一对一网络地址转换。NAT 服务器只拥有一个公有 IP 地址,某一时刻只允许最多一台内部主机访问外部网络。

(2)多对多网络地址转换。NAT 服务器拥有多个公有 IP 地址,可以满足多台内部主机并发访问外部网络的请求。

5.网络地址端口转换(NAPT)

按照普通的 NAT 方式,把一个内部主机的地址和一个外部地址一一映射后,在这条映射表项被清除之前,其他内部主机就不能映射到这个外部地址上。

NAPT(network address port translation,网络地址端口转换)是 NAT 的一种变形,它允许将多个内部地址映射到同一个公有地址上,达到多台内部主机同时访问外部网络的目的,可以显著提高公有 IP 地址的利用率。

NAPT 在映射 IP 地址的同时,还对传输层协议端口进行映射,不同的内部地址可以映射到同一个公有地址上,而它们的端口号被映射为该公有地址的不同端口号,亦即 NAPT 完成了"私有地址+端口"与"公有地址+端口"之间的转换。NAPT 有时也被称为 PAT 或 Address overloading。

6.内部服务器

NAT 隐藏了内部网络的结构,具有"屏蔽"内部主机的作用,但是在实际应用中,可能需要给外部网络提供一个访问内部主机的机会,如给外部网络提供一台 WWW 服务器,或一台 FTP 服务器。

7.Easy IP

Easy IP 是指进行地址转换时,直接使用接口的公有 IP 地址作为转换后的源地址。它也可以利用访问控制列表控制哪些内部地址可以进行地址转换。Easy IP 主要应用于将路由器 WAN 接口 IP 地址作为要被映射的公网 IP 地址的情形,特别适合小型局域网接入 Internet 的情况。一般具有以下特点:内部主机较少、出接口通过拨号方式获得临时(或固定)公网 IP 地址以供内部主机访问 Internet。

五、实验过程

1.实验任务一:配置 Basic NAT

在本实验中,实现私网主机 PCA 和 PCB 访问公网服务器 220.1.1.1,通过在 Router_2 上配置 Basic NAT,动态地为私网主机分配公网地址。

（1）搭建实验环境，初始化实验设备

按图 15-1 所示搭建实验环境，清除配置文件，以出厂配置重启网络设备。

```
<H3C>reset saved-configuration
<H3C>reboot
......
<H3C>
```

（2）基本配置

按图 15-1 所示，为 PCA 配置 IP 地址为 192.168.1.100/24，网关为 192.168.1.1；配置 PCB 的 IP 地址为 192.168.2.100/24，网关为 192.168.2.1；为 Router_1、Router_2、Router_3 各接口配置 IP 地址，并且确认各设备间直连网络能相互 Ping 通。IP 地址配完后在 Router_1、Router_2 静态配置私网内各网段的路由及默认路由，确保 PCA、PCB 能 Ping 通 200.1.1.1。为了给去往服务器的数据包提供路由，还需要在出口路由器 Router_2 上配置一条静态路由，指向 Router_3。

配置 Router_1:

```
[Router_1]interface GigabitEthernet 0/0
[Router_1-GigabitEthernet0/0]ip address 192.168.1.1 24
[Router_1]interface GigabitEthernet 0/1
[Router_1-GigabitEthernet0/1]ip address 192.168.2.1 24
[Router_1]interface Serial 6/0
[Router_1-Serial6/0]ip address 10.2.1.1 24
[Router_1]ip route-static 0.0.0.0 0 10.2.1.2
```

配置 Router_2:

```
[Router_2]interface Serial 6/0
[Router_2-Serial6/0]ip address 10.2.1.2 24
[Router_2]interface GigabitEthernet 0/0
[Router_2-GigabitEthernet0/0]ip address 200.1.1.1 24
[Router_2]ip route-static 192.168.0.0 16 10.2.1.1
[Router_2]ip route-static 0.0.0.0 0 200.1.1.2
```

配置 Router_3:

```
[Router_3]interface GigabitEthernet 0/0
[Router_3-GigabitEthernet0/0]ip address 200.1.1.2 24
[Router_3-GigabitEthernet0/0]quit
[Router_3]interface LoopBack 0
[Router_3-LoopBack0]ip address 220.1.1.1 32
[Router_3-LoopBack0]quit
[Router_3]ip route-static 200.1.1.0 24 200.1.1.1
```

在 PCA 上 Ping Router_2 公网的出口地址 200.1.1.1:

```
C:\Documents and Settings\Router__pca>ping 200.1.1.1
  Pinging 200.1.1.1 with 32 bytes of data:
    Reply from 200.1.1.1: bytes=32 time=20ms TTL=254
    Reply from 200.1.1.1: bytes=32 time=20ms TTL=254
```

```
    Reply from 200.1.1.1: bytes=32 time=20ms TTL=254
    Reply from 200.1.1.1: bytes=32 time=20ms TTL=254

Ping statistics for 200.1.1.1:
  Packets: Sent = 4, Received = 4, Lost = 0 (0% loss)
Approximate round trip times in milli-seconds:
Minimum = 20ms, Maximum = 20ms, Average = 20ms
```

在 PCB 上 Ping Router_2 公网的出口地址 200.1.1.1：

```
C:\Documents and Settings\Router__pcb>ping 200.1.1.1
 Pinging 200.1.1.1 with 32 bytes of data:
    Reply from 200.1.1.1: bytes=32 time=20ms TTL=254
    Reply from 200.1.1.1: bytes=32 time=20ms TTL=254
    Reply from 200.1.1.1: bytes=32 time=20ms TTL=254
    Reply from 200.1.1.1: bytes=32 time=20ms TTL=254

Ping statistics for 200.1.1.1:
      Packets: Sent = 4, Received = 4, Lost = 0 (0% loss)
Approximate round trip times in milli-seconds:
    Minimum = 20ms, Maximum = 20ms, Average = 20ms
```

（3）检查连通性

分别在 PCA，PCB 上 Ping 服务器,服务器 IP 地址为 220.1.1.1。

在 PCA 上 Ping 服务器：

```
C:\Documents and Settings\Router__pca>ping 220.1.1.1
 Pinging 220.1.1.1 with 32 bytes of data:
    Request timed out.
    Request timed out.
    Request timed out.
    Request timed out.

Ping statistics for 220.1.1.1:
Packets: Sent = 4, Received = 0, Lost = 4 (100% loss),
```

在 PCB 上 Ping 服务器：

```
C:\Documents and Settings\Router__pcb>ping 220.1.1.1
 Pinging 220.1.1.1 with 32 bytes of data:
    Request timed out.
    Request timed out.
    Request timed out.
    Request timed out.

Ping statistics for 220.1.1.1:
Packets: Sent = 4, Received = 0, Lost = 4 (100% loss),
```

结果显示，从 PCA，PCB 上无法 Ping 通服务器。这是因为私网地址在公网上无法被路由，从服务器回应的 Ping 报文在 Router_3 上无法找到 192.168.0.0/24 网段的路由。

（4）配置 Basic NAT

在 Router_2 上配置 Basic NAT。

通过 ACL 定义一条允许 192.168.0.0/16 网段的规则：

[Router_2]**acl number 2000**
[Router_2-acl-basic-2000]**rule permit source 192.168.0.0 0.0.255.255**
[Router_2-acl-basic-2000]**rule deny source any**

配置 NAT 地址池 1，地址的范围是 200.1.1.11 至 200.1.1.21。

[Router_2]**nat address-group 1 200.1.1.11 200.1.1.21**

进入接口模式，将地址池 1 与 ACL 2000 关联，应用到接口上方向为 outbound：

[Router_2]**interface GigabitEthernet 0/0**
[Router_2-GigabitEthernet0/0]**nat outbound 2000 address-group 1 no-pat**

（5）检查连通性

先用 PCA 上 Ping 服务器 IP 地址为 220.1.1.1，然后立即用 PCB Ping 服务器。显示如下：

```
C:\Documents and Settings\Router__pca>ping 220.1.1.1
Pinging 200.1.1.1 with 32 bytes of data:
    Reply rom 220.1.1.1: bytes=32 time=23ms TTL=253
    Reply from 220.1.1.1: bytes=32 time=21ms TTL=253
    Reply from 220.1.1.1: bytes=32 time=21ms TTL=253
    Reply from 220.1.1.1: bytes=32 time=21ms TTL=253

Ping statistics for 220.1.1.1:
Packets: Sent = 4, Received = 4, Lost = 0 (0% loss)
Approximate round trip times in milli-seconds:
Minimum = 21ms, Maximum = 23ms, Average = 21ms
```

在 PCB 上 Ping 服务器：

```
C:\Documents and Settings\Router_2>ping 220.1.1.1
 Pinging 220.1.1.1 with 32 bytes of data:
    Request timed out.
    Request timed out.
    Request timed out.
    Request timed out.

Ping statistics for 220.1.1.1:
Packets: Sent = 4, Received = 0, Lost = 4 (100% loss),
```

结果显示 PCA 能 Ping 通公网服务器，而 PCB 不能 Ping 通服务器。

（6）显示和调试 NAT 信息

用 PCA Ping 服务器后立即在 Router_2 上查看 NAT 表：

```
<Router_2>display nat session
There are currently 2 NAT sessions:
```

```
Protocol    GlobalAddr    Port    InsideAddr      Port    DestAddr    Port
-           200.1.1.11    ---     192.168.1.100   ---     ---         ---
VPN: 0,          status: NOPAT,       TTL: 00:04:00,      Left: 00:03:50
1           200.1.1.11    768     192.168.1.100   768     220.1.1.1   768
VPN: 0,          status: NOPAT,       TTL: 00:01:00,      Left: 00:00:50
```

从显示信息可以看出,该 ICMP 报文源地址 192.168.1.100 已经转换成公网地址 200.1.1.11,目的端口号和源端口号均为 768。

1min 后再次观察:

```
<Router_2>display nat session
There are currently 1 NAT session:
Protocol    GlobalAddr    Port    InsideAddr      Port    DestAddr    Port
-           200.1.1.11    ---     192.168.1.100   ---     ---         ---
VPN: 0,          status: NOPAT,       TTL: 00:04:00,      Left: 00:03:00
```

发现表中后两项消失了,4min 后再次观察:

```
[Router_2]display nat session
No NAT sessions are currently active!
```

从实验结果可以发现 4min 后所有表项都消失了。因为 NAT 表项具有一个老化时间(aging-time),一旦超过老化时间,NAT 会自动删除表项,前一条由系统为配置建立的 NAT 表项,老化时间为 4min,后一项五元组表项由流触发而建立,老化时间为 1min。可以通过 display nat aging-time 查看路由器的 NAT 默认老化时间。

```
[Router_2]display nat aging-time
NAT aging-time value information:
 tcp      ---- aging-time value is   86400 (seconds)
 udp      ---- aging-time value is     300 (seconds)
 icmp     ---- aging-time value is      60 (seconds)
 pptp     ---- aging-time value is   86400 (seconds)
 dns      ---- aging-time value is      60 (seconds)
 tcp-fin  ---- aging-time value is      60 (seconds)
 tcp-syn  ---- aging-time value is      60 (seconds)
 ftp-ctrl ---- aging-time value is    7200 (seconds)
 ftp-data ---- aging-time value is     300 (seconds)
```

通过命令 display address-group 查看地址池信息:

```
[Router_2]display nat address-group
NAT address-group information:
 1 : from    200.1.1.11   to    200.1.1.21
```

可以看到地址池的范围是 200.1.1.11 到 200.1.1.21。

(7)清除 Basic NAT 相关配置

删除 NAT 地址池:

```
[Router_2]undo nat address-group 1
Address-group is being used!
```

结果表明地址池的地址正在被使用,因此必须先删除 NAT 绑定。

在接口下删除 NAT 绑定:

```
[Router_2]interface GigabitEthernet 1/0
[Router_2-GigabitEthernet1/0]undo nat outbound 2000 address-group 1 no-pat
```

删除 NAT 地址池:

```
[Router_2]undo nat address-group 1
```

再用 display 查看地址池:

```
[Router_2]display nat address-group
NAT address-group information:
No address-groups have been configured
```

发现地址池已经成功被删除了。

2. 实验任务二:NAPT 配置

私网 PCA、PCB 需要访问公网服务器,但由于公网地址有限,可以通过配置 NAPT 动态地为 PCA、PCB 分配公网地址和协议端口。

(1)搭建实验环境,初始化实验设备

按图 15-1 所示搭建实验环境,清除配置文件,以出厂配置重启网络设备。

```
<H3C>reset saved-configuration
<H3C>reboot
```

(2)基本配置

按图 15-1 所示为 PCA 配置 IP 地址为 192.168.1.100/24,网关为 192.168.1.1;配置 PCB 的 IP 地址为 192.168.2.100/24,网关为 192.168.2.1;为 Router_1、Router_2、Router_3 各接口配置 IP 地址,并且确认各设备间直连网络能相互 Ping 通。IP 地址配完后在 Router_1、Router_2 静态配置私网内各网段的路由及默认路由,确保 PCA、PCB 能 Ping 通 200.1.1.1。为了给去往服务器的数据包提供路由,还需要在出口路由器 Router_2 上配置一条静态路由,指向 Router_3。

配置 Router_1:

```
[Router_1]interface GigabitEthernet0/0
[Router_1-GigabitEthernet0/0]ip address 192.168.1.1 24
[Router_1]interface GigabitEthernet 0/1
[Router_1-GigabitEthernet0/1]ip address 192.168.2.1 24
[Router_1]interface Serial 6/0
[Router_1-Serial6/0]ip address 10.2.1.1 24
[Router_1]ip route-static 0.0.0.0 0 10.2.1.2
```

配置 Router_2:

```
[Router_2]interface Serial 6/0
[Router_2-Serial6/0]ip address 10.2.1.2 24
[Router_2]interface GigabitEthernet 0/0
```

```
[Router_2-GigabitEthernet0/0]ip address 200.1.1.1 24
[Router_2]ip route-static 192.168.0.0 16 10.2.1.1
[Router_2]ip route-static 0.0.0.0 0 200.1.1.2
```

配置 Router_3：

```
[Router_3]interface GigabitEthernet 0/0
[Router_3-GigabitEthernet0/0]ip address 200.1.1.2 24
[Router_3-GigabitEthernet0/0]quit
[Router_3]interface LoopBack 0
[Router_3-LoopBack0]ip address 220.1.1.1 32
[Router_3-LoopBack0]quit
[Router_3]ip route-static 200.1.1.0 24 200.1.1.1
```

在 PCA 上 Ping Router_2 公网的出口地址 200.1.1.1：

```
C:\Documents and Settings\Router_pca>ping 200.1.1.1
Pinging 200.1.1.1 with 32 bytes of data:
  Reply from 200.1.1.1: bytes=32 time=20ms TTL=254
  Reply from 200.1.1.1: bytes=32 time=20ms TTL=254
  Reply from 200.1.1.1: bytes=32 time=20ms TTL=254
  Reply from 200.1.1.1: bytes=32 time=20ms TTL=254

Ping statistics for 200.1.1.1:
Packets: Sent = 4, Received = 4, Lost = 0 (0% loss)
Approximate round trip times in milli-seconds:
Minimum = 20ms, Maximum = 20ms, Average = 20ms
```

在 PCB 上 Ping Router_2 公网的出口地址 200.1.1.1：

```
C:\Documents and Settings\Router_pcb>ping 200.1.1.1
 Pinging 200.1.1.1 with 32 bytes of data:
    Reply from 200.1.1.1: bytes=32 time=20ms TTL=254
    Reply from 200.1.1.1: bytes=32 time=20ms TTL=254
    Reply from 200.1.1.1: bytes=32 time=20ms TTL=254
    Reply from 200.1.1.1: bytes=32 time=20ms TTL=254

Ping statistics for 200.1.1.1。
Packets: Sent = 4, Received = 4, Lost = 0 (0% loss)
Approximate round trip times in milli-seconds:
Minimum = 20ms, Maximum = 20ms, Average = 20ms
```

（3）检查连通性

分别在 PCA、PCB 上 Ping 服务器（IP 地址为：220.1.1.1）。
在 PCA 上 Ping 服务器：

```
C:\Documents and Settings\Router_pca>ping 220.1.1.1
Pinging 220.1.1.1 with 32 bytes of data:
   Request timed out.
   Request timed out.
```

```
  Request timed out.
  Request timed out.

Ping statistics for 220.1.1.1:
Packets: Sent = 4, Received = 0, Lost = 4 (100% loss),
```

在 PCB 上 Ping 服务器：

```
C:\Documents and Settings\Router__pcb>ping 220.1.1.1
Pinging 220.1.1.1 with 32 bytes of data:
  Request timed out.
  Request timed out.
  Request timed out.
  Request timed out.

Ping statistics for 220.1.1.1:
Packets: Sent = 4, Received = 0, Lost = 4 (100% loss),
```

结果显示，从 PCA，PCB 上无法 Ping 通服务器。

（4）配置 NAPT

在 Router_2 上配置 NAPT。
用 ACL 定义一条源地址属于 192.168.0.0/16 网段的流。

```
[Router_2]acl number 2000
[Router_2-acl-basic-2000]rule permit source 192.168.0.0 0.0.255.255
```

配置 NAT 地址池 1，地址池中只有一个地址 200.1.1.11

```
[Router_2]nat address-group 1 200.1.1.11 200.1.1.11
```

进入接口视图将 NAT 地址池 1 与 ACL 2000 绑定，接口方向为出方向。

```
[Router_2]interface GigabitEthernet 0/0
[Router_2-GigabitEthernet0/0]nat outbound 2000 address-group 1
```

此时命令中未携带 no-pat 关键字，表示 NAT 要对数据包进行端口转换。

（5）测试连通性

从 PCA、PCB 上 Ping 公网服务器。
在 PCA 上测试：

```
C:\Documents and Settings\Router__pca>ping 220.1.1.1
  Pinging 200.1.1.1 with 32 bytes of data:
    Reply from 220.1.1.1: bytes=32 time=21ms TTL=253
    Reply from 220.1.1.1: bytes=32 time=21ms TTL=253
    Reply from 220.1.1.1: bytes=32 time=21ms TTL=253
    Reply from 220.1.1.1: bytes=32 time=21ms TTL=253

Ping statistics for 220.1.1.1:
Packets: Sent = 4, Received = 4, Lost = 0 (0% loss)
Approximate round trip times in milli-seconds:
Minimum = 21ms, Maximum = 23ms, Average = 21ms
```

在 PCB 上测试：

```
C:\Documents and Settings\Router__pcb>ping 220.1.1.1
Pinging 200.1.1.1 with 32 bytes of data:
  Reply from 220.1.1.1: bytes=32 time=22ms TTL=253
  Reply from 220.1.1.1: bytes=32 time=21ms TTL=253
  Reply from 220.1.1.1: bytes=32 time=21ms TTL=253
  Reply from 220.1.1.1: bytes=32 time=21ms TTL=253

Ping statistics for 220.1.1.1:
Packets: Sent = 4, Received = 4, Lost = 0 (0% loss)
Approximate round trip times in milli-seconds:
Minimum = 21ms, Maximum = 23ms, Average = 21ms
```

发现 PCA、PCB 都能 Ping 通服务器。

（6）检查 NAT 表项

Ping 通后立即在 Router_2 上查看 NAT 表项：

```
[Router_2]display nat session
There are currently 2 NAT sessions:
Protocol    GlobalAddr      Port    InsideAddr      Port    DestAddr        Port
1           200.1.1.11      12289   192.168.2.100   43984   220.1.1.1       43984
VPN: 0,     status:     11,     TTL: 00:01:00,     Left: 00:00:51
1           200.1.1.11      12288   192.168.1.100   768     220.1.1.1       768
VPN: 0,     status:     11,     TTL: 00:01:00,     Left: 00:00:52
```

从表项中可以看出，源地址 192.168.1.100 和 192.168.2.100 转换成同一个公网地址 200.1.1.11，但端口号是不同的，192.168.1.100 转换的端口号为 768，而 192.168.2.100 转换的端口号为 43984，通过不同的端口号来分辨数据是发给 192.168.1.100 还是 192.168.2.100，NAPT 靠这种方式对数据包的 IP 层和传输层信息同时进行转换，从而显著提高 IP 地址的利用率。

（7）恢复配置

在 Router_2 上删除 NAPT 相关配置：

```
[Router_2]interface GigabitEthernet 0/0
[Router_2-GigabitEthernet0/0]undo nat outbound 2000 address-group 1
[Router_2-GigabitEthernet0/0]quit
[Router_2]undo nat address-group 1
```

3. 实验任务三：Easy IP 的配置

要实现私网内的主机访问服务器，但出口路由器 Router_2 上的出接口地址是动态获取的，因此需要在 Router_2 上配置 Easy IP。

（1）搭建实验环境，初始化实验设备

按图 15-1 所示搭建实验环境，清除配置文件，以出厂配置重启网络设备。

```
<H3C>reset saved-configuration
<H3C>reboot
```

（2）基本配置

按图 15-1 所示为 PCA 配置 IP 地址为 192.168.1.100/24，网关为 192.168.1.1；配置 PCB 的 IP 地址为 192.168.2.100/24，网关为 192.168.2.1；为 Router_1、Router_2、Router_3 各接口配置 IP 地址，并且确认各设备间直连网络能相互 Ping 通。IP 地址配完后在 Router_1、Router_2 静态配置私网内各网段的路由及默认路由，确保 PCA、PCB 能 Ping 通 200.1.1.1。为了给去往服务器的数据包提供路由，还需要在出口路由器 Router_2 上配置一条静态路由，指向 Router_3。

配置 Router_1：

[Router_1]**interface GigabitEthernet 0/0**
[Router_1-GigabitEthernet0/0]**ip address 192.168.1.1 24**
[Router_1]**interface GigabitEthernet 0/1**
[Router_1-GigabitEthernet0/1]**ip address 192.168.2.1 24**
[Router_1]**interface Serial 6/0**
[Router_1-Serial6/0]**ip address 10.2.1.1 24**
[Router_1]**ip route-static 0.0.0.0 0 10.2.1.2**

配置 Router_2：

[Router_2]**interface Serial 6/0**
[Router_2-Serial6/0]**ip address 10.2.1.2 24**
[Router_2]**interface GigabitEthernet 0/0**
[Router_2-GigabitEthernet0/0]**ip address 200.1.1.1 24**
[Router_2]**ip route-static 192.168.0.0 16 10.2.1.1**
[Router_2]**ip route-static 0.0.0.0 0 200.1.1.2**

配置 Router_3：

[Router_3]**interface GigabitEthernet 0/0**
[Router_3-GigabitEthernet0/0]**ip address 200.1.1.2 24**
[Router_3-GigabitEthernet0/0]**quit**
[Router_3]**interface LoopBack 0**
[Router_3-LoopBack0]**ip address 220.1.1.1 32**
[Router_3-LoopBack0]**quit**
[Router_3]**ip route-static 200.1.1.0 24 200.1.1.1**

在 PCA 上 Ping Router_2 公网的出口地址 200.1.1.1：

C:\Documents and Settings\Router__pca>**ping 200.1.1.1**
   Pinging 200.1.1.1 with 32 bytes of data:
      Reply from 200.1.1.1: bytes=32 time=20ms TTL=254
      Reply from 200.1.1.1: bytes=32 time=20ms TTL=254
      Reply from 200.1.1.1: bytes=32 time=20ms TTL=254
      Reply from 200.1.1.1: bytes=32 time=20ms TTL=254

Ping statistics for 200.1.1.1:
Packets: Sent = 4, Received = 4, Lost = 0 （0% loss）
Approximate round trip times in milli-seconds:
Minimum = 20ms, Maximum = 20ms, Average = 20ms

在 PCB 上 Ping Router_2 公网的出口地址 200.1.1.1：

```
C:\Documents and Settings\Router__pcb>ping 200.1.1.1
  Pinging 200.1.1.1 with 32 bytes of data:
    Reply from 200.1.1.1: bytes=32 time=20ms TTL=254
    Reply from 200.1.1.1: bytes=32 time=20ms TTL=254
    Reply from 200.1.1.1: bytes=32 time=20ms TTL=254
    Reply from 200.1.1.1: bytes=32 time=20ms TTL=254

Ping statistics for 200.1.1.1:
    Packets: Sent = 4, Received = 4, Lost = 0 (0% loss)
Approximate round trip times in milli-seconds:
    Minimum = 20ms, Maximum = 20ms, Average = 20ms
```

（3）检查连通性

分别在 PCA、PCB 上 Ping 服务器（IP 地址为 220.1.1.1）。

在 PCA 上 Ping 服务器：

```
C:\Documents and Settings\Router__pca>ping 220.1.1.1
Pinging 220.1.1.1 with 32 bytes of data:
    Request timed out.
    Request timed out.
    Request timed out.
    Request timed out.

Ping statistics for 220.1.1.1:
Packets: Sent = 4, Received = 0, Lost = 4 (100% loss),
```

在 PCB 上 Ping 服务器：

```
C:\Documents and Settings\Router__pcb>ping 220.1.1.1
  Pinging 220.1.1.1 with 32 bytes of data:
    Request timed out.
    Request timed out.
    Request timed out.
    Request timed out.

Ping statistics for 220.1.1.1:
Packets: Sent = 4, Received = 0, Lost = 4 (100% loss),
```

结果显示，从 PCA、PCB 上无法 Ping 通服务器。

（4）在 Router_2 上配置 Easy IP

通过 ACL 定义一条源地址为 192.168.0.0/24 网段的流。

[Router_2]**acl number 2000**
[Router_2-acl-basic-2000]**rule permit source 192.168.0.0 0.0.255.255**

在接口视图下将 ACL 2000 与其关联，方向为出方向。

[Router_2]**interface GigabitEthernet 0/0**

```
[Router_2-GigabitEthernet0/0]nat outbound 2000
```

（5）测试连通性

从 PCA、PCB 上 Ping 公网服务器。
在 PCA 上测试：

```
C:\Documents and Settings\Router_pca>ping 220.1.1.1
 Pinging 200.1.1.1 with 32 bytes of data:
    Reply from 220.1.1.1: bytes=32 time=21ms TTL=253
    Reply from 220.1.1.1: bytes=32 time=21ms TTL=253
    Reply from 220.1.1.1: bytes=32 time=21ms TTL=253
    Reply from 220.1.1.1: bytes=32 time=21ms TTL=253

Ping statistics for 220.1.1.1:
Packets: Sent = 4, Received = 4, Lost = 0（0% loss）
Approximate round trip times in milli-seconds:
Minimum = 21ms, Maximum = 23ms, Average = 21ms
```

在 PCB 上测试：

```
C:\Documents and Settings\Router_pcb>ping 220.1.1.1
 Pinging 200.1.1.1 with 32 bytes of data:
    Reply from 220.1.1.1: bytes=32 time=22ms TTL=253
    Reply from 220.1.1.1: bytes=32 time=21ms TTL=253
    Reply from 220.1.1.1: bytes=32 time=21ms TTL=253
    Reply from 220.1.1.1: bytes=32 time=21ms TTL=253

Ping statistics for 220.1.1.1:
Packets: Sent = 4, Received = 4, Lost = 0（0% loss）
Approximate round trip times in milli-seconds:
Minimum = 21ms, Maximum = 23ms, Average = 21ms
```

发现 PCA、PCB 都能 Ping 通服务器。

（6）检查 NAT 表项

```
[Router_2]display nat session
There are currently 2 NAT sessions:
Protocol   GlobalAddr   Port      InsideAddr      Port      DestAddr      Port
   1       200.1.1.1    12289     192.168.2.100   43985     220.1.1.1     43985
   VPN:  0,       status:    11,      TTL: 00:01:00,     Left: 00:00:43
   1       200.1.1.1    12288     192.168.1.100   768       220.1.1.1     768
   VPN:  0,       status:    11,      TTL: 00:01:00,     Left: 00:00:44
```

4. 实验任务四：配置 NAT Server

前面三个实验任务都实现了私网主机访问公网服务器，而且都是由私网主机主动发起连接的，如果公网要访问私网主机 PCA 的话，PCA 需要对外提供服务，需要在 Router_2 上配置 NAT Server。

（1）配置 NAT Server

在 Router_2 上配置 NAT Server：

[Router_2]**interface GigabitEthernet 0/0**
[Router_2-GigabitEthernet0/0]**nat server protocol icmp global 200.1.1.11 inside 192.168.1.100**

（2）检查连通性

从服务器上 Ping PCA 的公网地址 200.1.1.1。

```
[Router_3]ping 220.1.1.1
PING 200.1.1.11: 56  data bytes, press CTRL_C to break
    Reply from 200.1.1.11: bytes=56 Sequence=1 ttl=62 time=28 ms
    Reply from 200.1.1.11: bytes=56 Sequence=2 ttl=62 time=28 ms
    Reply from 200.1.1.11: bytes=56 Sequence=3 ttl=62 time=28 ms
    Reply from 200.1.1.11: bytes=56 Sequence=4 ttl=62 time=27 ms
    Reply from 200.1.1.11: bytes=56 Sequence=5 ttl=62 time=28 ms

  --- 200.1.1.11 ping statistics ---
    5 packet(s) transmitted
    5 packet(s) received
    0.00% packet loss
    round-trip min/avg/max = 27/27/28 ms
```

（3）查看 NAT Server

在 Router_2 上查看 NAT Server 表项：

```
[Router_2]display nat server
Server(s) in private network information:
        Currently 1 internal server(s) configured
Interface:Ethernet1/0, Protocol:1(icmp),
 [global]       200.1.1.11:      ----  [local]     192.168.1.100:       ----
```

表项信息中显示公网地址和私网地址一对一的映射关系，可以看出私网内的主机 192.168.1.100 使用公网地址 200.1.1.11 对外提供服务。

六、实验思考

① 在配置 Basic NAT 测试结果显示 PCA 能 Ping 通公网服务器，而 PCB 不能 Ping 通服务器，为什么？怎么解决？

② Basic NAT 和 NAPT 有何区别？各有什么优缺点？

③ Easy IP 在什么情况下使用比较合适？

实验十六 交换机端口安全技术

一、实验目的

① 掌握端口隔离的基本配置；
② 掌握端口绑定技术的基本配置。

二、实验内容

① 按组网图连接实验设备，进行连通性测试；
② 配置端口隔离，并进行验证；
③ 配置端口绑定，并进行验证。

三、实验设备与组网图

1．实验设备

二台 H3C0 交换机，一条专用 Console 配置线缆，四台计算机，超级终端或 Secure CRT 软件。

2．组网图

交换机端口安全技术实验组网如图 16-1。

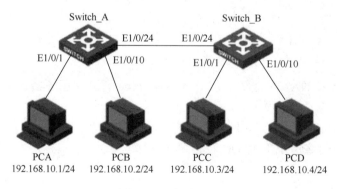

图 16-1 实验组网

四、实验相关知识

1. 端口隔离简介

端口隔离是为了实现报文之间的二层隔离,可以将不同的端口加入不同的 VLAN,但会浪费有限的 VLAN 资源。采用端口隔离特性可以实现同一 VLAN 内端口之间的隔离。用户只需要将端口加入到隔离组中,就可以实现隔离组内端口之间二层数据的隔离。端口隔离功能为用户提供了更安全、更灵活的组网方案。

端口隔离特性与端口所属的 VLAN 无关。对于属于不同 VLAN 的端口,只有同一个隔离组的普通端口到上行端口的二层报文可以单向通过,其他情况的端口二层数据是相互隔离的。对于属于同一 VLAN 的端口,隔离组内、外端口的二层数据互通的情况,又可分为以下两种。

① 支持上行端口的设备。
② 不支持上行端口的设备,隔离组内的端口和隔离组外端口二层流量双向互通。

2. 端口绑定技术简介

端口绑定是通过"MAC+IP+端口"绑定功能,可以实现设备对转发报文的过滤控制,提高了安全性。

五、实验过程

1. 实验任务一:配置隔离端口

(1)搭建实验环境,初始化实验设备

按图 16-1 所示搭建实验环境,清除配置文件,以出厂配置重启网络设备。

```
<H3C>reset saved-configuration
<H3C>reboot
```

(2)配置端口隔离

在交换机上启用端口隔离,设置端口 Ethernet 1/0/1、Ethernet 1/0/10 为隔离组的普通端口,Ethernet 1/0/24 为隔离组中的上行端口。

配置 Switch_A:

```
[Switch_A]interface Ethernet 1/0/1
[Switch_A-Ethernet1/0/1]port-isolate enable
[Switch_A-Ethernet1/0/1]quit
[Switch_A]interface Ethernet 1/0/10
[Switch_A-Ethernet1/0/10]port-isolate enable
[Switch_A-Ethernet1/0/10]quit
[Switch_A]interface Ethernet 1/0/24
```

```
[Switch_A-Ethernet1/0/24]port-isolate uplink-port
[Switch_A-Ethernet1/0/24]quit
```

（3）查看隔离组信息

```
[Switch_A]display port-isolate group
 Port-isolate group information:
 Uplink port support: YES
 Group ID: 1
 Uplink port: Ethernet1/0/24
 Group members:
    Ethernet1/0/1         Ethernet1/0/10
```

（4）端口隔离验证

在主机 PCA 上 Ping 主机 PCB，结果应该是不通，如下所示：

```
C:\Documents and Settings\Switch_pca>ping 192.168.10.2
Pinging 192.168.10.2 with 32 bytes of data:
 Request timed out.
 Request timed out.
 Request timed out.
 Request timed out.
……
```

在主机 PCA 上 Ping 主机 PCC，结果应该是通的，如下所示：

```
C:\Documents and Settings\Switch_pca>ping 192.168.10.3
Pinging 192.168.10.3 with 32 bytes of data:
 Reply from 192.168.10.3: bytes=32 time<1ms TTL=128
 Reply from 192.168.10.3: bytes=32 time<1ms TTL=128
 Reply from 192.168.10.3: bytes=32 time<1ms TTL=128
 Reply from 192.168.10.3: bytes=32 time<1ms TTL=128
……
```

以上结果表明，端口隔离配置成功。

2. 实验任务二：配置端口绑定

（1）配置端口绑定

在交换机上启用端口绑定。设置端口 Ethernet1/0/1 与 PCA 的 MAC 地址及 IP 地址绑定，端口 Ethernet1/0/10 与 PCB 的 MAC 地址及 IP 地址绑定。

配置 Switch_A：

```
[Switch_A]interface Ethernet 1/0/1
[Switch_A-Ethernet1/0/1]user-bind  ip-address  192.168.10.1  mac-address
0021-70FE-A294
[Switch_A-Ethernet1/0/1]quit
[Switch_A]interface Ethernet 1/0/10
[Switch_A-Ethernet1/0/10]user-bind  ip-address  192.168.10.2  mac-address
```

```
00D0-59CC-6053
```

（2）查看已设置的绑定信息

```
[Switch_A]display user-bind
Total entries found: 2
  MAC              IP              Vlan    Port                  Status
0021-70fe-a294    192.168.10.1     N/A     Ethernet1/0/1         Static
00d0-59cc-6053    192.168.10.2     N/A     Ethernet1/0/10        Static
```

（3）端口绑定验证

在主机 PCA 上 Ping 主机 PCB，其结果应该能通，如下所示：

```
C:\Documents and Settings\Switch_pca>ping 192.168.10.2
  Pinging 192.168.0.2 with 32 bytes of data:
    Reply from 192.168.10.2: bytes=32 time<1ms TTL=128
    Reply from 192.168.10.2: bytes=32 time<1ms TTL=128
    Reply from 192.168.10.2: bytes=32 time<1ms TTL=128
    Reply from 192.168.10.2: bytes=32 time<1ms TTL=128
    ……
```

断开 PCA、PCB 与交换机的连接，然后将 PCA 连接到端口 Ethernet 1/0/10，PCB 连接到端口 Ethernet 1/0/1。在重新用 Ping 来测试 PCA 到 PCB 的互通性，其结果应该是不通，如下所示：

```
C:\Documents and Settings\Switch_pca>ping 192.168.10.2
Pinging 192.168.10.2 with 32 bytes of data:
  Request timed out.
  Request timed out.
  Request timed out.
  Request timed out.
  ……
```

以上结果表示，端口绑定起了作用。

六、实验思考

端口隔离和端口绑定应该用在什么场合？用实例做设置说明。

综合实验一　网络设备调试与诊断操作

一、实验目的

① 掌握 IP 地址的使用和规划方法；
② 熟悉 Windows 操作系统常用网络管理命令；
③ 掌握 H3C 网络设备系统调试命令和诊断方法。

二、实验内容

① 按图综合 1-1 所示连接设备，配置主机和网络设备 IP 地址；
② 使用 Ping、Ipconfig、ARP、Netstat、Route、Tracert 等基本命令调试网络；
③ 学习 H3C 网络设备系统提供的调试诊断程序 debug 命令。

三、实验设备与组网图

1．实验设备

一台 H3C 路由器，一台交换机，三台计算机，四条双绞线，超级终端或 Secure CRT 软件。

2．组网图

网络设备调试与诊断操作如图综合 1-1 所示。

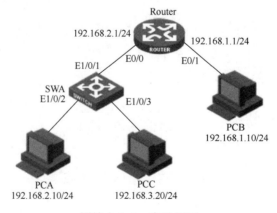

图综合 1-1　实验组网

四、实验相关知识

1．Windows 操作系统常用网络管理命令

（1）Ping 命令

Ping 命令是判断网络故障常用命令，是对两个 TCP/IP 系统连通性进行测试的基本工具，目的是测试目标主机是否可达。其命令格式如下：

```
ping [-t] [-a] [-n count] [-l length] [-f] [-i ttl] [-v tos] [-r count] [-s count]
[[-j computer-list] | [-k computer-list]] [-w timeout] destination-list
```

-t：Ping 指定的计算机直到中断。

-a：将地址解析为计算机名。

-n count：发送 count 指定的 ECHO 数据包数，默认值为 4。

-l length：发送包含由 length 指定的数据量的 ECHO 数据包。默认值为 32 字节；最大值是 65527。

-f：在数据包中发送"不要分段"标志。数据包就不会被路由上的网关分段。

-i ttl：将"生存时间"字段设置为 ttl 指定的值。

-v tos：将"服务类型"字段设置为 tos 指定的值。

-r count：在"记录路由"字段中记录传出和返回数据包的路由。count 可以指定最少 1 台，最多 9 台计算机。

-s count：指定 count 指定的跃点数的时间戳。

destination-list：指定要 Ping 的远程计算机。

（2）Ipconfig 命令

Ipconfig 是 Windows 操作系统中用于查看主机的 IP 配置命令，其显示信息中还包括主机网卡的 MAC 地址信息。该命令还可释放动态获得的 IP 地址，并启动新一次动态 IP 分配请求。

all：显示所有适配器的完整 TCP/IP 配置信息。在没有该参数的情况下，Ipconfig 只显示 IP 地址、子网掩码和各个适配器的默认网关值。

renew [adapter]：该参数在具有配置为自动获取 IP 地址的网卡的计算机上可用。更新所有适配器（如果未指定适配器），或特定适配器（如果包含了 adapter 参数）的 DHCP 配置。

release[adapter]：发送 DHCP release 消息到 DHCP 服务器，以释放所有适配器（如果未指定适配器）或特定适配器（如果包含了 adapter 参数）的当前 DHCP 配置并丢弃 IP 地址配置。该参数可以禁用配置为自动获取 IP 地址的适配器的 TCP/IP。

（3）ARP 命令

显示和修改 IP 地址与物理地址之间的转换表。

```
ARP -s inet_addr eth_addr [if_addr]
ARP -d inet_addr [if_addr]
```

```
ARP -a [inet_addr] [-N if_addr]
```

-s：添加主机，并将网络地址跟物理地址相对应，这一项是永久生效的。

-d：删除由 inet_addr 指定的主机，可以使用*来删除所有主机。

-a：显示当前的 ARP 信息，可以指定网络地址，不指定则显示所有的表项。

eth_addr：物理地址。

if_addr：网卡的 IP 地址。

inet_Addr：代表指定的 IP 地址。

（4）Netstat 命令

Netstat 可以显示路由表、实际的网络连接及每一个网络接口设备的状态信息。它用于显示与 IP、TCP、UDP 和 ICMP 协议相关的统计数据，一般用于检验本机各端口的网络连接情况。

netstat –r：显示路由表信息。

netstat –s：显示每个协议的状态，包括 TCP\UDP\ICMP 等。

netstat –n：以数字表格形式显示已经建立连接的 IP 地址和端口。

netstat –a：察看所有的连接。

（5）Route 命令

Route 命令是在本地 IP 路由表中显示和修改条目网络命令。一般使用 route delete、route add、route print 这三条命令可解决主机路由的所有功能。

route delete：删除路由。

route print：打印路由的 Destination。

route add：添加路由。

route change：更改现存路由。

（6）Tracert 命令

Tracert（跟踪路由）是路由跟踪实用程序，用于确定数据包到达目的主机所经过的路径，显示数据包经过的中继节点的清单和到达时间。Tracert 命令使用 IP 生存时间（TTL）字段和 ICMP 错误消息来确定从一个主机到网络上其他主机的路由。

其命令格式如下：

```
tracert [-d] [-h maximum_hops] [-j computer-list] [-w timeout] target_name
```

-d：指定不将地址解析为计算机名。

-h maximum_hops：指定搜索目标的最大跃点数。

-j host-list：与主机列表一起的松散源路由（仅适用于 IPv4）。

-w timeout：等待每个回复的超时时间（以毫秒为单位）。

-R：跟踪往返行程路径（仅适用于 IPv6）。

-S srcaddr：要使用的源地址（仅适用于 IPv6）。

-4：强制使用 IPv4。

-6：强制使用 IPv6。

target_name：目标计算机的名称。

2. H3C 网络设备系统调试命令

系统的命令行接口提供了种类丰富的调试功能，对于路由器所支持的各种协议和功能，基本上都提供了相应的调试功能，帮助用户对错误进行诊断和定位。

调试信息的输出可以由两个开关控制。

（1）terminal monitor

打开控制台对接口信息的监控功能。

（2）terminal debugging

屏幕输出开关，控制是否在某个用户屏幕上输出调试信息。

（3）debugging

协议调试开关，控制是否输出某协议的调试信息。

五、实验过程

1. 实验任务一：Windows 操作系统常用网络管理命令操作

（1）对设备进行基本的配置

1）将主机 PCA、PCB 的 IP 地址进行设置

如图综合 1-2 所示，将 PCA 主机地址设置为 192.168.2.10，子网掩码为 255.255.255.0，网关为 192.168.2.1；

将 PCB 主机地址设置为 192.168.1.10，子网掩码为 255.255.255.0，网关为 192.168.1.1。

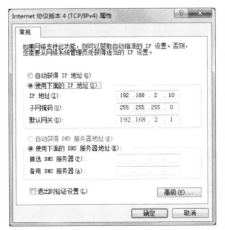

图综合 1-2　主机 PCA、PCB 配置

2）配置路由器

配置 E0/0 的 IP 地址为 192.168.2.1。

[Router]**int e0/0**
[Router-Ethernet0/0]**ip add 192.168.2.1 24**

配置 E0/1 的 IP 地址为 192.168.1.1。

[Router]**int e0/1**
[Router-Ethernet0/1]**ip add 192.168.1.1 24**

交换机不需要进行其他一些配置。通过这样的配置能够使 PCA 和 PCB 之间连通。

（2）通过 Ipconfig 命令查看本机网络配置信息

主机 PCA 打开"开始"菜单，选择"运行"，输入"cmd"命令，进入命令提示符窗口，如图综合 1-3 所示。

使用 Ipconfig 命令显示主机 PCA 适配器的完整 TCP/IP 配置信息。

```
G:\Documents and Settings\Administrator>ipconfig /all
```

图综合 1-3　主机 PCA 配置信息

（3）通过 Ping 命令测试连通性

主机 PCA 打开"开始"菜单，选择"运行"，输入"cmd"命令，进入命令提示符窗口。

① 使用 Ping 命令检查主机 PCA 与 192.168.2.1 网关的连通性。

```
G:\Documents and Settings\Administrator>ping 192.168.2.1
Pinging 192.168.2.1 with 32 bytes of data:
  Reply from 192.168.2.1: bytes=32 time<1ms TTL=255
  Reply from 192.168.2.1: bytes=32 time<1ms TTL=255
  Reply from 192.168.2.1: bytes=32 time<1ms TTL=255
  Reply from 192.168.2.1: bytes=32 time<1ms TTL=255

Ping statistics for 192.168.2.1:
    Packets: Sent = 4, Received = 4, Lost = 0 (0% loss),
Approximate round trip times in milli-seconds:
    Minimum = 0ms, Maximum = 0ms, Average = 0ms
```

"Reply from 192.168.2.1: bytes=32 time<1ms TTL=255"表示从 192.168.2.1 返回给主机

的数据包大小为 32 字节，时间小于 1ms，TTL 生存时间为 255ms。

"Packets: Sent = 4"可以看到主机向网关路由器发了 4 个 icmp 请求报文，"Received = 4"路由器返回给主机 4 个相应报文。"Lost = 0（0% loss）"，表明能够与路由器建立连通。

② 通过 Ping 命令检查 PCA 与 PCB 的网关的连通性

```
G:\Documents and Settings\Administrator>ping 192.168.1.1
  Pinging 192.168.1.1 with 32 bytes of data:
    Reply from 192.168.1.1: bytes=32 time=13ms TTL=255
    Reply from 192.168.1.1: bytes=32 time=4ms TTL=255
    Reply from 192.168.1.1: bytes=32 time=8ms TTL=255
    Reply from 192.168.1.1: bytes=32 time=8ms TTL=255

Ping statistics for 192.168.1.1:
Packets: Sent = 4, Received = 4, Lost = 0 (0% loss),
Approximate round trip times in milli-seconds:
Minimum = 4ms, Maximum = 13ms, Average = 8ms
```

表明主机能够与路由器的另一个接口，也就是 PCB 的网关进行通信。

③ 测试 PCA 与 PCB 的连通性。

```
G:\Documents and Settings\Administrator>ping 192.168.1.1
  Pinging 192.168.1.1 with 32 bytes of data:

    Reply from 192.168.1.1: bytes=32 time=13ms TTL=255
    Reply from 192.168.1.1: bytes=32 time=4ms TTL=255
    Reply from 192.168.1.1: bytes=32 time=8ms TTL=255
    Reply from 192.168.1.1: bytes=32 time=8ms TTL=255

Ping statistics for 192.168.1.1:
      Packets: Sent = 4, Received = 4, Lost = 0 (0% loss),
Approximate round trip times in milli-seconds:
      Minimum = 4ms, Maximum = 13ms, Average = 8ms
```

可以判断 PCA 与 PCB 能够建立连通性，它们之间能够进行通信。

（4）Tracert 命令应用

① 在 PCA 上 Tracert PCB。

```
G:\Documents and Settings\Administrator>tracert 192.168.1.10
Tracing route to 192.168.1.10 over a maximum of 30 hops
  1    <1 ms    <1 ms    <1 ms  192.168.2.1
  2    <1 ms    <1 ms    <1 ms  192.168.1.10
Trace complete.
```

② 在路由器上 Tracert PCB。

```
[H3C]tracert 192.168.1.10
  traceroute to 192.168.1.10（192.168.1.10）30 hops max,40 bytes packet,press CTR L_C to break
  1  192.168.1.10 1 ms 1 ms 1 ms
```

（5）Route 命令应用

如图综合 1-4 所示为主机 PCA 路由表。

`G:\Documents and Settings\Administrator>`**`route print`**

图综合 1-4　主机 PCA 路由表

2．实验任务二：IP 地址使用和规划

（1）步骤 1

① 设置两台主机的 IP 地址与子网掩码，两台主机均不设置缺省网关。

```
PCA: 192.168.2.10    255.255.254.0
PCC: 192.168.3.20    255.255.254.0
```

② 用 arp –d 命令清除两台主机上的 ARP 表。

`G:\Documents and Settings\Administrator>`**`arp -d`**

③ 然后在 PCA 与 PCC 上分别用 Ping 命令与对方通信，观察并记录结果，并分析原因。

`G:\Documents and Settings\Administrator>`**`ping 192.168.3.20`**

④ 在两台 PC 上分别执行 arp -a 命令，观察并记录结果。

提示：由于主机将各自通信目标的 IP 地址与自己的子网掩码相"与"后，发现目标主机与自己均位于同一网段（192.168.2.0），因此通过 ARP 协议获得对方的 MAC 地址，从而实现在同一网段内网络设备间的双向通信。

（2）步骤 2

① 将 PCA 的子网掩码改为 255.255.255.0，其他设置保持不变。

② 在两台 PC 上分别执行 arp -d 命令，清除两台主机上的 ARP 表。然后在 PCA 上"Ping" PCC，观察并记录结果。

G:\Documents and Settings\Administrator>**ping 192.168.3.20**

③ 在两台 PC 上分别执行 arp -a 命令，观察并记录结果。

提示：PCA 将目标设备的 IP 地址（192.168.3.20）和自己的子网掩码（255.255.255.0）相"与"，得 192.168.3.0，和自己不在同一网段，PCA 所在网段为 192.168.2.0，则 PCA 必须将该 IP 分组首先发向缺省网关。

（3）步骤 3

① 按照本实验任务步骤 2 的配置，接着在 PCC 上 Ping PCA，观察并记录结果。

G:\Documents and Settings\Administrator>**ping 192.168.2.10**

② 在 PCC 上执行 arp -a 命令，观察并记录结果。

3. 实验任务三：调试诊断程序 debug 命令操作

（1）诊断和定位

对于网络中的绝大部分协议和功能，系统都提供了相应的调试功能，帮助用户进行诊断和定位：

```
<H3C>terminal monitor
% Current terminal monitor is on
```

打开控制台对接口信息的监控功能：

```
<H3C>terminal debugging
% Current terminal debugging is on
```

打开调试信息的屏幕输出开关。在调试之前必须打开这两个开关，调试信息才会在终端显示出来。可以通过"？"来查看哪些协议能够被调试。

```
<H3C>debugging ?
  all               All debugging functions
  arp               ARP module
  aspf              ASPF module
  atm               ATM module
  bgp               BGP protocol
  bridge            Debugging bridge
  connection-limit  Connection-limit module
  dar               Specify DAR configuration information
  device            Device managed
  dhcp              Dynamic Host Configuration Protocol
  dialer            DCC module
  dns               Domain name system module
  dot1x             802.1x configuration information
  encrypt-card      Specify Encrypt-card configuration information
  ethernet          Ethernet module
  fib               FIB module
  firewall          Firewall module
  fr                FR module
```

```
   ftp-server      FTP server information
   gre             Generic Routing Encapsulation
   http            Hypertext transfer protocol
   hwtacacs        Specify HWTACACS server
   ifnet           Turn on IFNET switch
---- More ----
```

Debug 命令使用的时候要很谨慎，尽量给出 Debug 监控具体细节，因为如果 Debug 要显示的信息量太大，会导致机器立刻瘫痪。

（2）在 PCA 中 Ping 路由器的 192.168.1.1 接口

查看路由器的显示信息：

```
<H3C> debugging
*Apr  2 16:44:21:899 2010 H3C IPDBG/7/debug_icmp:
ICMP Receive: echo(Type=8, Code=0), Src = 192.168.2.10, Dst = 192.168.1.1
*Apr  2 16:44:21:899 2010 H3C IPDBG/7/debug_icmp:
ICMP Send: echo-reply(Type=0, Code=0),
Src = 192.168.1.1, Dst = 192.168.2.10
*Apr  2 16:44:22:900 2010 H3C IPDBG/7/debug_icmp:
ICMP Receive: echo(Type=8, Code=0),
Src = 192.168.2.10, Dst = 192.168.1.1
*Apr  2 16:44:22:900 2010 H3C IPDBG/7/debug_icmp:
ICMP Send: echo-reply(Type=0, Code=0),
Src = 192.168.1.1, Dst = 192.168.2.10
*Apr  2 16:44:23:900 2010 H3C IPDBG/7/debug_icmp:
ICMP Receive: echo(Type=8, Code=0),
Src = 192.168.2.10, Dst = 192.168.1.1
*Apr  2 16:44:23:900 2010 H3C IPDBG/7/debug_icmp:
ICMP Send: echo-reply(Type=0, Code=0),
Src = 192.168.1.1, Dst = 192.168.2.10
*Apr  2 16:44:24:900 2010 H3C IPDBG/7/debug_icmp:
ICMP Receive: echo(Type=8, Code=0),
Src = 192.168.2.10, Dst = 192.168.1.1
*Apr  2 16:44:24:900 2010 H3C IPDBG/7/debug_icmp:
ICMP Send: echo-reply(Type=0, Code=0),
Src = 192.168.1.1, Dst = 192.168.2.10
```

可以看到 Ping 命令是有去有回的。当主机发送了一个 ICMP 报文后，路由器会对这个报文进行响应，发送回一个响应报文。所以通过 Debug 可以查看到 8 个 ICMP 报文。从报文中还可以看出报文的源 IP 地址和目的 IP 地址，是 ICMP 请求报文还是响应报文。通过这样的信息能够判断网络当中遇到的一些问题。

六、实验思考

① 试分析实验任务二中 arp -d 命令的作用。
② 试分析实验任务二中步骤 3 的结果。

综合实验二 网络互联技术综合应用实践

一、实验任务

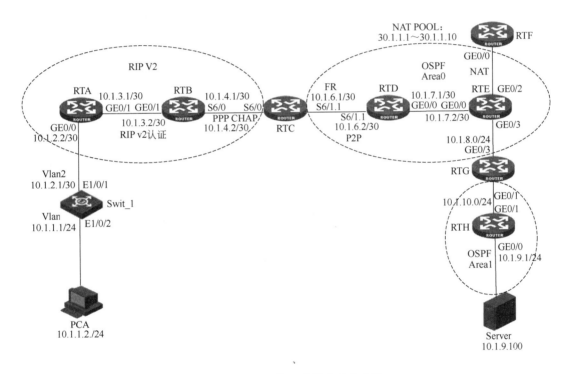

图综合 2-1 网络互联技术综合实验

二、实验要求

按图综合 2-1 所示连接设备，对路由器和交换进行基本配置，实现互联互通。
① 在交换机上创建 Vlan，给端口配 VLAN。
② 对路由器 RTA 作适当配置，实现不同 VLAN 互通。
③ RTA 和 RTB 之间进行 RIPv2 认证。
④ RTB 和 RTC 之间实现 PPP（CHAP）验证。
⑤ RTC 和 RTD 之间实现 FR 的配置。

⑥ 在路由器 Router_B 上设置访问 Internet 的缺省路由。

⑦ 在路由器 RTC 上进行 RIPv2 和 OSPF 的相互注入，保证网络连通。

⑧ 在 RTE 上实现 NAT 转换。

⑨ 进行适当的 ACL 设置，使 PCA 能访问 Server 的 Web 和 Email 服务，但不能进行 FTP 访问。

参 考 文 献

［1］ 杭州华三通信技术有限公司．构建中小企业网络 H3CNE V6.0 培训教材（内部资料）．
［2］ 杭州华三通信技术有限公司．构建 H3C 高性能园区网络学习指导书（上册）（内部资料）．
［3］ 杭州华三通信技术有限公司．构建 H3C 高性能园区网络学习指导书（下册）（内部资料）．
［4］ 杭州华三通信技术有限公司．构建 H3C 高性能园区网络实验指导书（内部资料）．
［5］ 杭州华三通信技术有限公司．H3C 大规模网络路由技术学习指导书（内部资料）．
［6］ 杭州华三通信技术有限公司．H3C 大规模网络路由技术实验指导书（内部资料）．
［7］ 杭州华三通信技术有限公司．构建安全优化的广域网学习指导书（上册）（内部资料）．
［8］ 杭州华三通信技术有限公司．构建安全优化的广域网学习指导书（下册）（内部资料）．
［9］ 杭州华三通信技术有限公司．构建安全优化的广域网实验指导书（内部资料）．
［10］杭州华三通信技术有限公司．路由交换技术：第 1 卷（上册）［M］．北京：清华大学出版社，2011．
［11］杭州华三通信技术有限公司．路由交换技术：第 1 卷（下册）［M］．北京：清华大学出版社，2011．
［12］杭州华三通信技术有限公司．路由交换技术：第 2 卷［M］．北京：清华大学出版社，2011．
［13］杭州华三通信技术有限公司．路由交换技术：第 3 卷［M］．北京：清华大学出版社，2011．
［14］杭州华三通信技术有限公司．路由交换技术：第 4 卷［M］．北京：清华大学出版社，2011．
［15］特南鲍姆，著．计算机网络［M］．第 5 版．严伟，潘爱民，译．北京：清华大学出版社，2012．
［16］雷震甲．网络工程师教程［M］．第 4 版．北京：清华大学出版社，2014．
［17］张纯容，施晓秋，刘军．网络互连技术［M］．北京：电子工业出版社，2015．
［18］Stevens W R．TCP/IP 详解 卷 1：协议［M］．范建华，译．北京：机械工业出版社，2011．
［19］Stevens W R．TCP/IP 详解 卷 2：实现［M］．范建华，译．北京：机械工业出版社，2011．